The Woodburner's Companion

The Woodburner's Companion

Practical Ways of Heating with Wood

by Dirk Thomas

Foreword by Castle Freeman

Alan C. Hood & Company, Inc.
CHAMBERSBURG, PENNSYLVANIA

Illustrated by Carl Kirkpatrick and Stacey Ann Boughrum

Designed by James F. Brisson
Typeset by Jessica L. Clark

ISBN 0-911469-20-6

Published by Alan C. Hood & Company, Inc.
Chambersburg, PA 17201

10 9 8 7 6 5 4 3 2 1

Library of Congress Cataloging-in-Publication Data

Thomas, Dirk, 1946-
    The woodburner's companion : practical ways of heating with wood / Dirk
Thomas.
        p. cm.
    Includes bibliographical references (p.123).
    ISBN 0-911469-20-6
        1. Stoves, Wood. 2. Fuelwood. 3. Dwellings--Heating and ventilation. I. Title

TH7437 .T46 2001
697'.22--dc21

                                                                                    00-053841

CONTENTS

FOREWORD

Burning wood for home heat is one of those activities—like bird watching, playing the piano, and banking your wages—in which you can go exactly as far as you want to go and still get the benefit. As the author of this engaging and practical book suggests, you can burn three logs in a fireplace on Christmas Eve and call it quits for the year; or you can build a fire in a woodburning appliance of some kind on Columbus Day and keep it going until Memorial Day, by-passing all other sources of heat. Wherever you find yourself on that long, long scale, you will be helped by this book to burn wood economically, efficiently and—most important—safely.

That is because *The Woodburner's Companion* is not difficult material—or not exactly. Burning wood is easy. Most people have done it at one time or another, in camping, say, or cooking out, or in yard work.* Indeed, to the householder today who feels himself burdened by higher and higher fuel bills, heating with wood may seem like the easiest, most natural thing in the world. He's right, too, but there's a catch. Woodburning is like driving a car: easy only as long as you do it more or less right, otherwise an invitation to degrees of failure, expense and misfortune. Therefore, if our householder is to get some significant part of his home's heat from wood without wasting vast amounts of time and money—and without burning his house down—he needs a little guidance.

Enter *The Woodburner's Companion*. Its author, Dirk Thomas, writes from the State of Vermont, the heart of the heart of the wood-

---

*So it is that every American male over the age of fifteen thinks he knows how to build and tend a fire in a fireplace. Many of them are wrong. The book at hand can help them, too. See Chapter 4.

burning belt. He has a lifetime's experience as a chimney sweep, firewood cutter and woodburner in those parts. He knows every kind of wood heater there is, from eighteenth-century open fireplaces to modern catalytic stoves that amount to the woodburning equivalent of the space shuttle. He knows wood itself, as a fuel: how to get it, how to keep it, how to use it. He also knows the factors that must go right if you are to burn wood successfully and safely; and he knows what has probably gone wrong in woodburning systems that make more trouble than comfort. Most remarkably, the author writes about what he knows in a way that is clear, simple and thoroughly likable.

In the end, of course, you learn how to heat your home successfully with wood mainly through experience, something for which no book can be a substitute. What the best instructional books can do is to keep you out of the worst kinds of trouble as you go about acquiring experience, and so help you to live and learn. *The Woodburner's Companion* can fill that role admirably. If you doubt that, if the excellence of Dirk Thomas's knowhow and experience, wedded to his ability as a writer, seem to you too much for his publisher to claim, read his book and see for yourself.

Castle Freeman, Jr.
Newfane, Vermont
October 2000

INTRODUCTION

Why a book about wood heat on the eve of the 21st Century? Isn't wood an antiquated fuel? Aren't there far more convenient ways of staying warm?

My answer is that, yes, wood is an antiquated fuel—antiquated in the sense that it has served mankind well for centuries—but it is accessible regardless of Middle Eastern (or American) politics and crashing computers. And, yes, wood is an inconvenient heat source, but for people who learn to accommodate the inconvenience, it's a heat source that offers far more than warmth.

As I write this, petroleum prices for the coming winter are forecast to sharply increase, and there is growing testimony from oil company geologists that *real*, not manipulated, shortages are not very far down the road.

Meanwhile, wood heating equipment has improved drastically since the 1970's. Wood may not be *the* fuel of the future, but it is certainly *a* fuel of the future for millions of people weary of high-priced fossil fuels, power outages and possible shortages. It is my hope that this book will prove useful to those considering wood heat as well as to those who already enjoy it.

If you are among those who have read my earlier book, *The Harrowsmith Country Life Guide to Wood Heat*, you will quickly realize that this book is, largely, an updated version of the earlier work. Some parts were dropped and some were added, but a liberal amount of pilferage occurred. Nevertheless, I hope this book will introduce wood heating to a new generation of readers, and so make its own way. I'm told that a British bartender would likely say, if you were to order a second pint, "Similar, sir, not the same."

The
Woodburner's
Companion

CHAPTER 1

Fuel of the Future?

Woodburning has been a source of fascination for mankind since prehistory. To consider how it has accompanied us on our journey through time is to conjure up colorful and disparate pictures: heavy-browed hunters holding their hands to a warming fire at the mouth of their cave; tipis (or wagons) circled in the prairie night, darkness held at bay by the glowing fire in the center of the ring; the yule log blazing on the hearth of a medieval castle; the glowing cookstove warming the remote New England farmhouse in the depths of an old-fashioned January. Heat is a necessity—it's also a talisman against the dark, a symbol of home, safety, celebration and self-reliance.

It is true that during a period of recent history, wood heating dwindled in most industrialized nations. You'll find few ads for woodstoves in the popular periodicals of the '50s and '60s. Rural folk still burned wood, of course, but it took the Arab oil boycott of the '70s to reacquaint many of us with our almost lost heritage. The woodburning revival that followed, fueled as it was by soaring oil prices and apparently dwindling supplies of the petroleum products we had taken for granted, had an air of desperation. It was as though we were collectively faced with the task of reinventing the wheel on the very day that we needed to use the wagon. Mistakes were made: houses designed for central heating were now using wood-fired space heaters; people whose only knowledge of home heating consisted of turning up the thermostat when the weather got cold were quite suddenly attempting to cut their own wood and install and operate wood heaters; woodstoves were being designed and built by people whose only qualifications were ownership of a welding torch and a grasp of the dubious principle that bigger is better. The results were predictable: a lot of trees that shouldn't have been were cut and burned; a lot of chimneys puffed a lot of particle-laden smoke and suffered chimney fires; some houses burned; and some people got injured or killed in the fires or in woodcutting accidents. But also, millions of North American households freed themselves partially or entirely from dependence upon imported petroleum products for residential heat, and a new industry was born: the loose infrastructure of stove and fireplace dealers, manufacturers, installers and researchers, chimney sweeps and firewood producers that serves today's woodburner.

As the '70s gave way to the '80s and oil prices lost their shock value, a certain number of people returned to the convenience of oil, gas or electric heat. Millions of others, however, having discovered that they liked the savings, the nature of the heat, the independence or all three, kept on burning wood. Most recently, the large increase in conventional energy prices during the winter of 1999-2000 seems to be spurring an equivalent increase in interest in wood heat.

Thanks to the twenty-plus years of research spurred by the woodburning revival of the '70's, we know a great deal more than we did about wood, stoves, chimneys and wood heat in general. Thanks in part to Environmental Protection Agency and Canadian Provincial regulations, we have fewer stove manufacturers than in the '70's, but the stoves they produce are far cleaner burning and more effective than were the airtight boxes of two decades ago. Nevertheless, the future of wood heat is largely dependent upon two questions that must be conclusively answered: 1.) Is wood heat economically viable? And, most important, 2.) Is wood heat environmentally defensible?

The final answers—as final as answers to this sort of question can be—will come partly from researchers and manufacturers, but it is, in large part, the individual woodburner who holds the key. Will we be educated and responsible enough to make full use of technological advances in wood heating? Or will we cut the wrong trees and burn their wood improperly in the wrong stoves until laws springing from environmental necessity force us to abandon our woodpiles in favor of the oil tank, the gas pipeline, the electric meter?

Bearing in mind that wood heat and people who burn wood don't lend themselves very well to generalizations, let us attempt to find preliminary answers to our two questions.

Is Wood Heat Economically Viable?

I'm reminded of a tongue-in-cheek article I saw in the '70's that purported to be an accounting of the "real" expenses involved in cutting your own winter's supply of firewood. Included were a $10,000 truck, chain saws, trips to the hospital emergency room and much lost time at work. With these costs figured in, it's needless to say, the cost per cord of a winter's fuel was prohibitive. There can be a certain amount of truth to this gloomy view of cutting your own wood; for some people, it's a good way to save money, but it's not for every woodburner, particularly since wood can be purchased cut, split and delivered in most areas at prices that make it a bargain when compared with other fuels.

Simple math shows that, other things being equal, you will spend fewer dollars heating your house with wood than you will with oil, gas, coal or electricity. A cord of good hardwood will produce as much heat as will 200 gallons of heating oil, and the comparison with other fuels is similar. What this means, in dollars and cents, is that if oil costs $1 per gallon, you could pay $200 for a cord of wood—an unheard of price outside of major cities—and be getting as much heat for your money. So why doesn't everyone use wood? Because, harking back to the apocryphal tale with which we began this discussion, simple math does not address all of the factors that the prospective woodburner must consider. Your time, for example. Heating with wood is a decidedly active pursuit; even if you buy your wood ready to burn, you'll devote considerable time to piling, carrying, stoking and ash removal. If you cut your own, you'll find yourself warmed more than twice by the felling, bucking, splitting and hauling you'll do, and you'll also find that the time you commit to getting in your wood makes heating your home more of an ongoing hobby than an occasional chore.

Does it make economic sense, then, to heat your home with wood? Yes, if you have more time than money, and yes, if you enjoy the work and ritual unique to this form of heat. Burning wood fits some ways of life in the same way that vegetable gardening and livestock raising do: it also saves money, but the savings are almost incidental to the satisfaction it can provide.

Is Wood Heat Environmentally Defensible?

There are two major environmental issues connected to wood heat: air pollution and deforestation.

The air pollution issue has been something of a cause celebre in the wood heat industry. Some municipalities now have strictly enforced "no-burn" days and limits on installations of woodburning appliances, and the Environmental Protection Agency has mandated that, as of 1990, no stoves may be manufactured that emit more than 4.1 grams of particulate matter per hour. (To put that figure in perspective, consider that the "black box" stoves typical of the 1970's poured up to 40 grams per hour into the atmosphere.)

Woodsmoke contains carcinogens, so the concern is legitimate. Other environmental alarms about woodsmoke require closer examination. Concerning the "Greenhouse Effect," for example, in which carbon dioxide in woodsmoke degrades the Earth's ozone layer leading to global warming, it is worth noting that the carbon dioxide produced by burning wood would have been produced even had the wood not been burned, when the trees that it came from died and decomposed. (This is not true of fossil fuels, however, since they are extracted from below the Earth's surface and do not decompose.)

The response of the wood heat industry to environmental concerns is that the new, high-tech stoves, with their drastically reduced emission levels (some models now produce less than 1 gram per hour), will vastly reduce air pollution from woodburning. Also, there has been some suspicion in the industry that wood heat has been unfairly singled out as a culprit. A story that made the rounds of chimney sweeps a while ago concerned a community with a no-burn day. The story had it that the inspector whose job it was to drive around and order people to extinguish their fires would have had to make a bust every ten or fifteen minutes to prevent more pollution than his car was putting into the atmosphere. This may be true, but in any event it is clearly incumbent upon the woodburning public to do its share toward alleviating the potentially catastrophic problems of air pollution and ozone depletion by learning to burn wood properly and by switching to cleaner burning stoves whenever possible.

The issue of deforestation does not lend itself to simple answers. While it is true that firewood harvesting can be an excellent tool for encouraging sound forest management, it is equally true that not all firewood is harvested responsibly. It is safe to say, however, that massive deforestation—in tropical and temperate rainforests, for example—is associated with timber production and land clearing, not firewood production. Vermont, as a case in point, is a bastion of woodburning but still grows more wood every year than it cuts.

A last point concerning the ethics of woodburning is that no form of home heating, with the possible exception of solar, is environmentally benign. And the nature of wood as heating fuel makes it less harmful than other fuels in certain ways not immediately related to the burning of the fuel. Consider:

Carrying. Wood, since it is generally produced locally, requires far less transportation than other heat sources—no oil tankers, pipelines, long-haul trucks, electrical transmission lines, etc.

Accidents. A spilled load of firewood is an inconvenience, not an ecological disaster.

Supply. Responsibly harvested, wood is a renewable energy source.

Politics. Wars over firewood have been rare or nonexistent, and there is no reason to believe that this will change. Woodburners are mostly too busy cutting, splitting and piling to do much fighting.

Environmental impact. Electricity and natural gas require substantial rights-of-way for transmission and distribution. These are, in effect, clearcuts which must be perpetually maintained, often with herbicides.

Toxins. Neither wood nor its residue—ashes—is a hazardous waste.

I could go on, but I think I've made my point: Every kind of fuel, including wood, has an impact on the environment. Is heating with wood ethical? It may not be if you live in an area plagued by air pollution. It probably isn't if you live in an area with little forested land. It isn't if you harvest and burn your wood irresponsibly. If, on the other hand, your circumstances permit and you decide to become a responsible user of the resource, woodburning can be an integral part of a contained and conserving way of life with positive ecological impacts balancing the negative. It is my intention that this book be a useful tool for aiding and encouraging responsible, ethical and enjoyable use of firewood.

Who Should Burn Wood?

Who should burn wood? This looks like a simple question, but considering the time, money and frustration (not to mention the discomfort) that may attend a wrong answer, it's worth a fairly exhaustive investigation. What we are really asking is not one simple question, but a series of involved ones.

Where do you live? In a valley where atmospheric inversions cause frequent smog? In an area with a relatively warm climate? A place with little nearby woodland? We previously discussed the increasing awareness on the part of the public and certain governmental regulatory agencies that woodsmoke, for all its evocative and benign associations, is an air pollutant—an important one in some areas. Having already stated that I feel wood heating is getting a somewhat undeserved environmental bum rap born of oversimplification, I must admit that this most ancient and cozy source of heat can have negative effects.

I remember when it first occurred to me that wood heating was potentially harmful. It was probably 1975 or 1976—the heart of the woodburning revival and the era of the airtight black-box stove—and I was driving into Rutland, Vermont, at dawn. It was a bitterly cold and clear January morning, and as I topped the hill that afforded me my first view of the little city, I was shocked to see what appeared to be a deep lake of cloud hovering over the narrow Rutland valley, supported by countless plumes of chimney smoke. Vermont's air then—as it does most of the time now—seemed wonderfully clean and invigorating to me, but this dense cloud was not alien, not a bad-mannered tourist from Newark or Los Angeles; it was our very own, a creation of our innocent desire to warm ourselves with the resources at hand. Oh, I know, car exhaust and oil burner fumes contributed to that cloud, but the distinct and pervasive odor of woodsmoke in Rutland that morning left no doubt about what was largely to blame.

The smog that I saw that cold morning was partly a result of poorly designed and operated stoves, but mainly a result of the concentration of a large number of those stoves on the floor of a fairly steep-walled valley. It's worth noting that none of the surrounding hill towns had smog that morning, even though many of their houses were assuredly heated by poorly designed and operated stoves. The culprit, then, was an atmospheric inversion, defined as cold air at ground level, warm air above, and no mixing or air movement occurring, because cold air doesn't rise. Inversion most often occurs in valleys—particularly steep-walled valleys—because they collect and hold cold air.

The lesson to be taken from this? If you live in a well-populated valley, you'd be well advised to think twice before planning your winter around a woodstove: besides contributing to a real air pollution problem, you may—now or in the future—find your use of wood heat seriously constrained by local ordinances. At the very least, plan on using a new-generation, low-emission stove.

How cold are your winters? Inversion-prone areas aren't the only places unsuitable for serious wood heating. If you live in a warm climate—winter temperatures seldom going below freezing—you may find that a woodstove causes more problems than it solves. As a chimney sweep in northern New England, I have learned that the dirtiest and therefore most dangerous chimneys belong to people who run their stoves regularly in the spring and autumn. They light a fire to take the chill of a cool rainy day out of the house, but once the stove has accomplished that, it keeps throwing heat, forcing the owner to either shut off its air supply and cool the stove down or go outside and stand in the rain. Woodstoves—both black boxes and low-emission units—don't burn efficiently at very low temperatures; smoldering fires produce large quantities of air pollution and creosote at a cost to the environment and

your safety that outweighs any savings in fuel expenses you might realize. If winters where you live aren't cold enough to consistently demand relatively high-temperature burning, you might do better with an
alternative heat source.

Can you get wood easily? The local availability of firewood is another
factor to consider. Two of the most important tangible advantages
that wood has over other heating fuels are that it's relatively cheap and
plentiful and—a related factor—it requires relatively low-impact transportation. If you live in an area with few trees—the Great Plains, for
example—or in an area whose trees are not suitable for harvest—
Montreal—wood's advantages as a fuel become purely theoretical. Truck-
ing wood long distances increases the price dramatically and eliminates
or decreases one of wood's environmental advantages over, say, oil:
low impact, short-haul transportation. Cutting trees for fuel in areas
with little forestland is selfishly short-sighted. This doesn't mean that
you'd have to live in the middle of the forest primeval to burn wood
with a clear conscience, just that fuel wood should be a by-product of
sound forestry practices in your locality. Before you decide to become
a serious fuel wood consumer, investigate the availability of wood in
your area.

How is your home laid out? If the topography, climate and sylvan
population of your area check out, the next factor to evaluate is your
house. Unhappy unions of heaters with homes are commonplace, often troublesome and potentially dangerous. The design and siting of
your house may dictate what sort of woodburning appliance—if any—
you should use, but before considering this, let's talk a bit about wood
heaters and the types of heat that they produce.

Wood heaters fall into three broad categories: Stoves, fireplaces
(which aren't normally "serious" heaters) and furnaces. Stoves and fireplaces produce radiant heat, which travels in straight lines and heats
solid objects in its path rather than the air. Furnaces, which may be no
more than stoves installed in the basement and connected to ductwork
or water pipes, as well as some specially designed stoves, produce primarily circulating heat—warming the air or water which transfers its
heat as it is circulated. Neither the radiant nor the circulating heater has
an intrinsic advantage over the other, but your house may make one
preferable.

If your house, for example, is a typical ranch—spread out but not up, with numerous interior walls, the most reasonable way to safely heat it with wood is with a furnace in the basement. On the other hand, if part of your reason for considering wood heat is esthetic, and a stove that you can sit in front of is what you want, then resign yourself to heating part of your home, but not all of it, with wood. There's nothing wrong with this—in fact, many wood heating experts feel that the smaller stoves appropriate for this sort of space heating are easier to operate efficiently than are the appliances large enough to heat a whole house. People do heat entire ranch houses with radiant woodstoves, but this usually means using two or more small stoves or keeping the room where the large stove is located unbearably hot.

A house designed or renovated for the purpose can be nicely heated with a woodstove. Such a house is probably compact, multistoried to take advantage of rising heat, and with relatively few interior walls to inhibit the spreading of the stove's radiant heat. We should note that many such houses are also sited to maximize solar gain. This is a wonderful feature, as I can attest since my house is one of these, but it has an impact on the wood heating operation. To illustrate, I'll use myself as an example. For some years I had a stove with a thermostatically controlled air supply; the stove would automatically close its air inlet when the room got warm. This meant that if I guessed the weather wrong and stoked the stove on the morning of a day that became sunny, I'd come home to find that my heating system had shut itself down to a slow burn and thus had been distilling creosote all day. Even a nonthermostatic stove in a solar house can cause problems if you've stoked it full of wood and the sun comes out. It may not shut itself down, but it will surely provoke you to do so, and that, of course, will cause rapid creosote buildup. Open some windows instead, and memorize those fair weather signs.

Is there such a thing as a house which, because of its design, should not be at least partially heated with wood? Perhaps. I'd propose as a candidate a well-insulated, very tightly constructed one-storied house with no basement (thus making a wood furnace impractical) and no large rooms—a ranch-style house built on a slab, for instance, or a mobile home. (Note also that woodburning appliances in mobile homes may be subject to Housing and Urban Development Department re-

strictions.) In such a house, it would be nearly impossible to burn a
stove hot enough for efficiency without overheating the room where it
was located. The one feasible option would be an outdoor boiler: the
wood-fired appliance is outside and is plumbed to provide hot water to
the house for central heating. Outdoor boilers clearly have attractive
features, but they have significant disadvantages as well. We'll discuss
them more fully later.

In all likelihood, your house is neither designed expressly for wood
heat nor completely impractical for it, which means that, perhaps with
some renovations, you could use wood to fill part or all of your heating
needs. In subsequent chapters, we will discuss how in detail.

Do you have the time required? The final factor to consider in deciding
whether you should burn wood is you. Are you willing to take the time
to learn how to do it properly and then follow through? Are you willing to deal with wood, ashes, dust and the inconvenience of a source of
heat which, though lovely, is not automatic? Wood may save you money,
but it won't save you time, and its pleasures make up for its drawbacks
only if you feel those pleasures keenly. Many people, of course, heat
with wood out of absolute financial necessity; and a number of them,
though by no means all, would drop it like a bad habit if they could
afford to. For those of you who have the option, I encourage you to
choose wood, but only after carefully weighing all of the considerations
just indicated.

CHAPTER 2

The Heating Value of Wood

O ne of the most appealing things about wood—to me, at least—is the lore that surrounds it: how long must you season wood? Should you mix green and dry wood? Does wood split more easily when it's green? People who burn wood actually discuss (argue about) these and many other questions. Gas furnaces may be convenient, but when was the last time you heard people arguing about them?

In any event, there is a great deal to be learned concerning the heating value of various kinds of wood—usually as compared to other kinds of wood, but also, by extension, as compared to other fuels. One would suppose that simple measurements would determine heat values for wood, but over the years I've seen many tables purporting to show these values, and few of these tables have been in total agreement. Science, it appears, hasn't universally displaced lore.

I suspect that much of the variation in the tables can be explained by the nature of wood. Trees, like woodburners, are individuals. The available Btu content of one cord of "air- dried" white ash may not be the same as that of another, because "air-dried" is a scientifically nebulous term: it tells you only how the wood was stored, not what its moisture content—and, therefore, useful heat value—actually is. Perhaps one tree grew on a wetter site and had a higher initial moisture content. Perhaps it was piled uncovered and didn't get as dry.

A problem with tables which compare wood to other fuels is that, to be truly accurate, they must account for the differences in efficiency among the various appliances in which various fuels are burned. The heat value of a cord of hickory may be equivalent to that of a couple of hundred gallons of fuel oil, but this is purely theoretical if you burn the hickory in an open fireplace, since much of the heat will go up the chimney. The good tables do factor appliance efficiencies into their comparisons, but even this is problematic: the efficiencies established for woodstoves in a laboratory setting may bear scant resemblance to how those same stoves perform in your home. Also, the tables I've seen predate the highly efficient new stoves and therefore probably assign too low an efficiency rating to stoves in general. To help rectify this problem, I'm including a table from Natural Resources Canada that gives up-to-date efficiency ratings for most types of heaters.

Before we get to the heating value table, a few things to consider: the "best" wood may vary with the situation. For example, shagbark hickory has the highest Btu content among North American woods, but some woodburners complain that it "coals up" excessively and doesn't burn actively enough to keep a house warm in very cold weather. (Solve this problem by mixing the hickory with other species. Stop complaining.) Or suppose it's rainy, 35 degrees Fahrenheit—dreary, in short—and you want a fire for both spiritual and physical warmth. Is white oak (22,700,000 Btu's per cord) or white birch (18,900,000 Btu's) the "best" wood?

Because your situation is unique, only experience can give you answers to most such questions. But rest assured that experience has the power to do just that.

Now for the tables. The first one gives the heat content of a number of common varieties of wood and compares them with coal, fuel oil and natural gas. This table comes from a pamphlet called "Wood as a Home Fuel," published by the Cooperative Extension Services of the Northeast States. I've chosen this table because it includes a large number of wood species, and because the heat values it assigns are in agreement with a majority of the other tables I've seen. The second table, from "A Guide to Residential Wood Heating," published by Natural Resources Canada, gives up-to-date efficiency ratings for common residential heaters.

TABLE 1

Heat Values of Various Wood Species Compared to Other Fuels

GREATEST HEAT EQUIVALENTS
1 Cord = 1 Ton of Anthracite Coal (Approx.)

A Wood *(1 standard cord)*	B *Available heat of* *1 cord. wood (Btu's)*	C *Anthracite coal* *(tons)*	D *No.2 fuel oil* *(gallons)*	E *Natural gas* *(100 cu. ft.)*
Hickory, shagbark	24,600,000	1.12	251	308
Locust, black	24,600,000	1.12	251	307
Hophornbeam	24,100,000	1.09	246	301
Apple	23,877,000	1.09	244	298
Elm, rock (or cork)	23,488,000	1.07	240	294
Hickory, bitternut	23,477,000	1.07	240	293
Oak, white	22,700,000	1.04	232	284
Beech, American	21,800,000	.99	222	273

HIGH HEAT EQUIVALENTS
1 Cord = 9/10 Ton of Anthracite Coal (Approx.)

A Wood *(1 standard cord)*	B *Available heat of* *1 cord. wood (Btu's)*	C *Anthracite coal* *(tons)*	D *No.2 fuel oil* *(gallons)*	E *Natural gas* *(100 cu. ft.)*
Birch, yellow	21,300,000	.97	217	286
Maple, sugar	21,300,000	.97	217	286
Oak, red	21,300,000	.97	217	286
Ash, white	20,000,000	.91	204	250
Walnut, black	19,500,000	.89	198	244
Birch, white	18,900,000	.86	193	236
Cherry, black	18,770,000	.85	191	235

14 THE WOODBURNER'S COMPANION

MODERATE HEAT EQUIVALENTS
1 cord = 8/10 Ton of Anthracite Coal (Approx.)

A Wood (1 standard cord)	B Available heat of 1 cord. wood (Btu's)	C Anthracite coal (tons)	D No.2 fuel oil (gallons)	E Natural gas (100 cu. ft.)
Tamarack	18,650,000	.85	190	233
Maple, red	18,600,000	.84	190	232
Ash, green	18,360,000	.83	187	229
Pine, pitch	17,970,000	.82	183	225
Sycamore	17,950,000	.82	183	224
Ash, black	17,300,000	.79	177	216
Elm, American	17,200,000	.78	176	215
Maple, silver	17,000,000	.77	173	213

LOW HEAT EQUIVALENTS
1 Cord = 6/10 Ton of Anthracite Coal (Approx.)

A Wood (1 standard cord)	B Available heat of 1 cord. wood (Btu's)	C Anthracite coal (tons)	D No.2 fuel oil (gallons)	E Natural gas (100 cu. ft.)
Spruce,red	13,632,000	.62	139	170
Hemlock	13,500,000	.61	138	169
Butternut	12,800,000	.58	131	160
Pine, red	12,765,000	.58	130	160
Aspen (poplar)	12,500.000	.57	128	156
Pine, white	12,022,000	.55	123	150
Basswood	11,700,000	.53	119	146
Fir, balsam	11,282,000	.51	115	141

A. 1 standard cord =128 cubic feet wood and air; 80 cubic feet solid wood; 20% moisture content. 1 lb. of this wood contains 5,780 Btu's (British thermal unit). (1 Btu=amount of heat required to raise temperature of a pound of water 1 degree Fahrenheit.)

B. It is assumed that available heat is measured based on oven-dry, minus heat loss due to moisture, minus loss due to water vapor formed, minus loss due to heat carried away in dry chimney gas. Stack temperature 450 degrees F. No excess air. Efficiency of burning unit=50 to 60%.

C. Contains 28,000,000 Btu's per ton, but available heat is only 22,000,000 Btu's per ton. 1 lb. of coal contains 11,000 available Btu's. Coal burned under similar conditions to wood.

D. 1 gallon contains 140,000 Btu's, but is burned at 70% efficiency, providing 98,000 available Btu's.

E. 100 cu. ft.=1 therm=100,000 Btu's, but is burned at 80% efficiency, providing 80,000 available Btu.

SOURCE: *Wood as a Home Fuel* published by the Cooperative Extension Services of the Northeast States.

TABLE 2

Typical Heating System Efficiencies

Fuel	Type of System	a Efficiency
Oil	Conventional Burner	60%
	Retention Head Burner	70—78%
	Advanced Mid-efficiency furnace	83—89%
Electricity	Central Furnace or Baseboard	95%
Natural Gas	Central Furnace—conventional	55—65%
	—powered exhaust	75—82%
	—condensing	88—96%
Propane	Central Furnace—conventional	55—65%
	—powered exhaust	76—83%
	—condensing	85—93%
Wood	Central Furnace	45—55%
	Conventional Stove (properly located)	55—70%
	"High Tech" Stove (properly located)	70—80%
Wood Pellets	Pellet Stove	55—80%

a. Efficency = percentage of Btu's in fuel which become
"sensible" heat in the living space.

SOURCE: *A Guide to Residential Wood Heating* published by National
Resources Canada.

CHAPTER 3

Strategies for Using Wood

Some people burn wood once a year—in an open fireplace on Christmas Eve—and some people spend the entire six months of a northern winter feeding five or ten cords of wood to the constantly burning stove or furnace that is all that stands between them and freezing. Many other people take—or would like to take—a middle road between occasional ambience creation and total reliance. To them wood is a supplement to the fossil fuel burner, reducing their dependence upon oil or gas. Or perhaps they keep a woodstove hooked up in the event of a power outage. Just two winters ago an ice storm knocked out the power for well over a week in parts of Quebec, Vermont and New York State. A good many people in the affected area had woodstoves, and a good many others subsequently got them.

More recently, a sharp increase in the price of conventional heating fuels has made many people wonder if the age of cheap and plentiful petroleum which we have enjoyed for many years and is drawing to a close. Woodstove retailers and installers report substantial new and renewed interest in the oldest home heating fuel.

Which of the possible uses of fuel wood best suits you? If you really are only interested in the holiday fire, you've probably already stopped reading, so I'll assume that your intentions go beyond the purely recreational.

In picking your strategy, you should consider the factors discussed earlier—availability of wood, your house, etc.—since these may limit your options. If, for example, you live in a metropolitan area where firewood is expensive and storage room limited, you can't reasonably consider doing all of your heating with wood.

You must also consider the way you live. Heating with wood is a hands-on process; if nobody is home twelve hours a day, then wood should not be your sole heating fuel. Let's consider some wood heating strategies.

Wood as the Sole Source of Heat

I suspect that a relatively small percentage of households are suited to total reliance on wood. For one thing, as discussed before, your house must be appropriate for—if not designed around—wood heat. For another, you must live in an area where wood is abundant and available. You must be ready, willing and able each year to handle an awful lot of wood (figure on three to five cords for a smallish newer house in a cold climate). And you must forgo long absences during heating season, since most stoves and furnaces require attention every eight hours at the least.

Recognize also that wood heat is problematic in the spring and fall when you need heat, but not very much. Drastically reducing the air supply to a stove in an attempt to control its heat output will likely create uncceptable amounts of creosote in your chimney and pollution in your (and your neighbors') air. Unless somebody is home often enough to kindle several small fires a day, you'll either be too warm, too cold, or too polluting in marginal heating weather.

Finally, from personal experience, I can assure you that handling firewood does not get easier as you grow older, and any number of common infirmities can render it totally impractical. With heat, as with most important matters, it seems wise to have several baskets for your eggs.

Wood as the Primary Source of Heat

This strategy will fit a few more households than the first: you've got a back-up heating system, so you have more flexibility. Your stove won't completely run your life six months a year, but you'll probably burn nearly as much wood as will those in the first category, so your house, location and lifestyle need to be nearly as accommodating.

Wood as Supplementary Heat

Even if your circumstances make major reliance on wood heat impractical, you might find that a stove or fireplace stove which heats part of your house some of the time will give you substantial savings on your fuel bills and a good deal of pleasure and comfort in the bargain. Such a part-time heater will probably do the most good on both counts if you are able to locate it in the most-used common space of your house. Note that properly sizing a wood heater to the area it will be heating is critically important, particularly if that area is only a room or two; a large stove properly operated will make a small area unbearably hot. Knowledgeable stove dealers and chimney sweeps can help you pick a stove appropriate to your situation.

Making a point of using your part-time stove during cold spells will probably enhance its value to you; you will likely discover, as many have, that when sub-zero air is finding all of the leaks in your walls and windows, nothing feels warmer than a woodstove.

Wood as an Emergency Back-up Heat Source

I mentioned the catastrophic ice storm of 1998 earlier, but even the lesser power outages to which rural areas are prone can be uncomfortable or even dangerous in severe weather. A just-in-case woodstove and a small supply of wood can turn a wretched situation into a merely inconvenient one.

A Caveat

It's far from true—distressingly so—that all people who rely heavily on wood heat have properly installed, operated and maintained heating systems, but I fear that there is even more of a tendency toward negligence among occasional woodburners. Even if you only use your stove on weekends or during power failures, an unsafe installation or a dirty chimney poses a serious risk to you and your home. All wood heating systems should be installed, used and maintained to the highest standards of safety.

CHAPTER 4

Equipment and Techniques

It's time to ponder variables: As noted in Chapter 1, wood heat is available in three basic forms: stoves, furnaces (in which group I include boilers) and fireplaces. In addition, each main group has subgroups, and there are several interesting heaters—pellet stoves, masonry heaters and fireplace stoves—that don't neatly fit in any of the three categories. To determine which choice is best for you, consider the components of your heating system, potential or existing: you, your house, your chimney, your location and the heating appliances under consideration.

Whether you choose a stove, fireplace, furnace or masonry heater, the rest of your system will affect and, to varying degrees, be affected by the decision. For example, as we saw in an earlier chapter, the design of your house (*e.g.,* small, very well insulated with lots of interior partitions) may make a woodstove impractical unless you undertake large-scale renovations.

Your first step, therefore, is to consider what you want from wood heat: esthetic enjoyment? An open fireplace is probably the purest embodiment of that. Lower fuel bills? The more of your heating requirement that is met by burning wood, the more money you will save. A stove or furnace can sharply reduce or eliminate your dependence on other heating fuels. Or do you want both the esthetic charm of woodburning and the savings? You might lean toward a wood-fired boiler in your basement and a fireplace in your living room, or you might find that a handsome woodstove with a viewing window or a high-tech fireplace stove will satisfy both needs. The point is to have a wood heating system that does what you want it to do and works the way it is supposed to—a system in which all parts are well matched. To that end, I'll discuss the different types of wood heaters with an eye to their strengths, weaknesses and best applications.

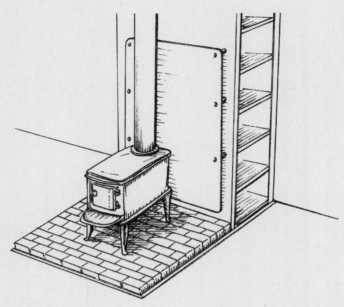

Woodstoves

Woodstoves have certainly changed. The 1990 advent of the EPA's final stage of emissions standards has brought us cleaner burning, more efficient and more expensive stoves. Their place in the woodburning spectrum remains unchanged, however: a stove is a space heater capable, under some circumstances, of heating an entire house and providing, for most users, some esthetic satisfaction.

Among stoves there are four subgroups: old airtight, older nonairtight, catalytic and high-tech noncatalytic.

The old airtights—the "black boxes" of the 1970s—are still available secondhand, as people upgrade to the more efficient modern models. A good airtight can do a very effective job of heating and will generally hold a fire overnight, but will produce dangerous amounts of creosote and unconscionable quantities of air pollution if burned incorrectly.

Older, nonairtight stoves—most of them antiques at this point—are less readily available, but not impossible to find. Often ornate, these stoves, because their combustion air supply is hard to control, tend to produce less creosote than do airtights, but because so much of the heat that they produce goes up the chimney, they use more wood than a modern stove will to heat the same space, and cannot be counted upon to hold a fire overnight.

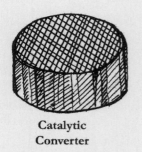

Catalytic Converter

Catalytic stoves employ a catalytic converter or combustor to reduce air pollution, with the happy side-effect of producing more heat per cord of wood. The converter, a ceramic honeycomb coated with either platinum or palladium, reduces the ignition temperature of the tars, vapors and other organic compounds that make up woodsmoke. When the converter is pre-heated to light-off temperature (500-700 degrees Fahrenheit) and the gate that allows the smoke to bypass the catalyst during preheating is closed, the catalyst itself burns much of the smoke. The result is that creosote formation is reduced by as much as 90%. In addition, because of the extra heat the burning smoke produces, wood use is frequently reduced by one-third. Catalytic stoves generally hold their fires as long as airtights, but they require the operator to pay attention to properly preheating (but not overheating) the catalyst, and they need new catalysts—often at over $100 a pop—after ten to twelve thousand hours (two to four seasons of regular burning).

The noncatalytic high-tech stoves rely upon secondary combustion instead of catalytic converters to achieve the clean burn mandated by the EPA and Canadian Provincial regulations. They attain secondary combustion by guiding the smoke coming off the burning wood to hot spots in the firebox and, usually, by introducing a fresh supply of combustion air, with the result that temperatures in excess of 1,000 degrees Fahrenheit burn the smoke. High-tech stoves do not burn quite as cleanly as catalytic stoves working at peak efficiency, but they may well do so over the course of several seasons, since the cat stoves' performance probably declines near the end of their converters' lifespan. High-tech stoves, like catalytic stoves, hold fires overnight and require that the operator pay attention to maintaining the correct temperature, particularly at the beginning of the burn cycle.

BEST APPLICATIONS

Stoves are not central heating equipment, so they should be located where the heat is most needed—in the living space. If your house was designed to be heated by a stove or has, by fortuitous circumstance, few interior partitions and is compact rather than sprawling, you likely can do all or almost all of your heating with a woodstove. Most houses don't lend themselves to this, however; as observed earlier, a stove large enough to heat a good-sized ranch-style house would probably make the room in which it was located unbearably hot. The same is true of almost any house without an open floor plan, so for most people, the most realistic role for a woodstove is as supplemental heat.

ADVANTAGES

Price. Although new stoves typically cost $1,000-$2,000, they can amortize themselves rapidly. Corning, Inc., manufacturers of catalytic converters used in stoves, once calculated that heating a 2,000-square-foot house with a catalytic stove will cost $325 less per season than heating the same house with an oil furnace. As oil prices increase, so will the savings.

Looks. A furnace represents woodburning's utilitarian extreme—so ugly that it's hidden in the basement, but it heats the whole house—and an open fireplace represents the esthetic extreme: beautiful and mesmerizing, but, in many houses, so inefficient that it results in a net heat loss. Woodstoves reside between the extremes: you may not be able to heat your entire house with one, and the stove may not provide the ambience of an open hearth, but it fulfills both needs quite well. For 12 winters I've enjoyed the company of my soapstone stove, which has an iron frame with the walls, top and floor made of soapstone. It has a large viewing window, so I can see the flames, and it heats the entire house. It seems to me to be a very good compromise.

Adaptability. Installing a woodstove in a house is rarely as easy as it might seem, but it usually is less expensive, disruptive and time-consuming than installing a furnace or fireplace.

DISADVANTAGES

Mess. Woodstoves are messy. Dust, sawdust and bits of bark constantly try to take up residence in the vicinity of a stove, leaving you three alternatives: live with the mess; devise a way to minimize the mess and deal with it easily (woodbox near the stove, hearth broom and shovel handy); or be driven progressively more crazy by it until you give up the stove. I've seen all three solutions practiced, and recommend the second, but I suspect that true neat freaks will find themselves dreading the heating season if they have a woodstove.

Dry House Syndrome. I believe I just coined this phrase. Winter air, being cold, contains low amounts of moisture to begin with, and all forms of wood heat, with the exception of wood-fired steam heat systems, are dry and exacerbate the problem. This leads to dry skin and throats, failure of glued furniture joints and great suffering on the part of wooden musical instruments. Although dry house syndrome is a problem with most forms of wood heat, I'm hanging it on woodstoves because they are generally running constantly, and because of their usual location: right where the skin, throats, chairs and mandolins and pianos are.

Work. Stoves require active, knowledgeable participation by their owners. So do fireplaces and furnaces, of course, but perhaps not as much. I don't consider this a weakness, by the way, at least not until the end of March, but I don't doubt that people who feel strongly that heat should be a convenience will disagree. Stoves are nothing if not inconvenient. It's not, I hasten to add, that you must constantly baby and coddle a woodstove, but neither can you simply set a thermostat and allow anonymous technology to do the rest.

Fireplace Stoves

This relatively new wood heating option is really a form of woodstove, but deserves separate discussion because of its esthetic qualities and because of its installation possibilities. In the 1970's, lots of people, lacking a proper separate flue, installed woodstoves in their fireplaces simply by shoving the stovepipe up the chimney. These usually horrible installations are not what I refer to in this category. Modern fireplace stoves are highly efficient units designed to convert a conventional fireplace into a useful heater while sacrificing little, if any, of its charm. There are also fire-

place stoves designed to be installed alone—that is, without a pre-existing fireplace. These devices have the appearance of a fireplace with glass doors, but provide the heating power of a woodstove.

BEST APPLICATIONS

Fireplace stoves provide as much, or nearly as much, heat as a high-tech or catalytic stove and, as mentioned above, they look very much like fireplaces, so they are appropriate for people who value both heating power and esthetics too much to sacrifice either. In fairness to free-standing stoves, many of the newer models have large viewing windows, and are handsome in their own right, but if you have a bad case of fireplace lust, they still look like stoves; a Vermont Castings Winterwarm, for example, does not. Additionally, people with conventional masonry fireplaces who wish to convert them to serious heaters may find that the fireplace stove is the best answer.

ADVANTAGES

Looks. As discussed above, a nice fireplace stove is certainly handsome enough to be the centerpiece of a room.

Efficiency. Since these units have the same technological features as other modern stoves, you can expect to get a lot of heating bang for your buck. Their only disadvantage in this regard is that, because they are either in an existing fireplace or built into a wall, they have only one side open to your living space and, thus, tend to release their heat more slowly. Many of them have built-in fans which largely overcome this problem.

DISADVANTAGES

Price. The fireplace stoves which are not installed in existing fireplaces require the construction of either a mass-insulated metal chimney (a double-wall chimney with the space between the walls filled with insulating material) or a triple-wall metal chimney. This might add $2,000 to the bill, and since the stoves themselves cost $2,000 or more, you are looking at a considerable investment. The units that are installed in existing fireplaces won't save you much, because it is necessary (by law in Canada, by common sense only, so far, in the U.S.) to reline the flue, since fireplace flues are almost universally too large and cold for modern stoves, and in order to provide an unbroken direct connection from stove to chimney top, both for the sake of draft and in order to avoid removing the stove from the fireplace opening for cleaning.

Mess and work. See above, since fireplace stoves are stoves in every way but appearance.

Furnaces and Boilers

It is technically incorrect to lump furnaces, which heat by circulating warm air, with boilers, which circulate heated water. From the homeowner's standpoint, however, the appliances have much in common: they provide central heat, they are not located in the living area, and aspects of their operation, such as control of combustion air, are typically more automated than are a woodstove's. I have seen conventional basement-installed woodstoves that use the oil furnace's ductwork to circulate warm air through the house, and I suppose that the addition of a plenum to gather heated air transforms these units into primitive furnaces, but what I'm describing here are appliances designed specifically to be central heaters. (*Note:* The primitive furnace just described may be very unsafe, since wood-fired appliances generally produce higher temperatures than do oil or gas-fired units. Your pre-existing ductwork may not have adequate clearance from combustible parts of your house to handle the additional heat.)

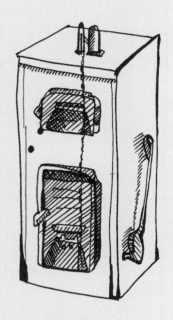

As to which is best, furnace or boiler, each has its advocates. A furnace may be cheaper to install, but a boiler heats water, a more efficient heat-transfer medium than air. Furthermore, a hot-water system doesn't blow air into the living space, an advantage, since moving air feels cooler than still air of the same temperature. In the end, your choice in an existing house will likely be based upon the oil or gas-fired system in place: if it's forced hot air, you'll probably be loath to install expensive new plumbing. Another cautionary note: the installation of a wood-fired boiler or furnace, even if it's an add-on (one that uses the existing heat circulation system), should only be done by a thoroughly knowledgeable professional. Incorrectly installed furnaces and boilers may function poorly and can be subject to dangerous heat and/or pressure buildup. Also, the woodburning appliance should not share a chimney flue with the oil or gas-fired appliance unless it is designed and approved for that sort of installation.

BEST APPLICATIONS

Large or sprawling houses that cannot be effectively heated with a stove usually can be with a furnace or boiler. Also, furnaces and boilers don't require a lot of operation. You do have to stoke them, but modern units pretty much do the rest (except ash removal) based on the thermostat setting. That relative convenience makes furnaces and boilers appealing to those who seek the economic benefits of woodburning but who don't feel the need for a woodstove as the centerpiece of the living area.

ADVANTAGES

Location. Furnaces and boilers are usually in the basement, so that's where the mess associated with woodburning will be confined (unless you have to carry wood through the house to get it to the basement—a situation that I expect you'd quickly rectify).

Effectiveness. Properly installed, a wood furnace or boiler will heat an entire house fairly evenly, unlike a woodstove.

Convenience, already discussed.

DISADVANTAGES

Price. Furnaces and boilers are generally much more expensive than stoves—several thousand dollars and up, not counting installation or any ductwork or plumbing that may be necessary. If your house can be heated with a stove, it's unlikely that the furnace will ever pay for itself by comparison. If your only alternative is a fossil-fuel-fired appliance, however, the wood-fired model will amortize itself in time.

Diffuse heat. The disadvantage of central heating is the disadvantage of the wood furnace or boiler: there's no place to go to get warm, unless you care to set up a rocking chair in your basement. You wouldn't in mine.

Dependence on electricity. Most furnaces and boilers need electricity to work, since they depend upon forced drafts, pumps or blowers. This could be a problem if you are subject to frequent power failures.

Lack of charm. I've never seen a beautiful furnace or boiler: handsome in a utilitarian way, perhaps, but never soothing to gaze upon.

Relative lack of efficiency. Woodburning furnace and boiler technology has not kept pace with that of woodstoves; they seem to occupy the same shadowy (perhaps smoky and smoggy as well) world as SUV's: the pollution that they cause is officially ignored. The result is that furnaces and boilers generally use more wood and create more creosote and pollution than do woodstoves. I doubt that this is due to negligence on the part of manufacturers: it is likely the nature of the beast. A furnace is, of necessity, large, and will be capable of producing much more heat than is needed in all but the most severe weather. More importantly, the air supply is adjusted in response to a thermostat setting, rather than in response to the condition of the fire. The net result is frequent low-temperature burning which, as we've discussed, is highly inefficient.

Outdoor Boilers

Outdoor boilers are an increasingly popular form of wood-fired central heat, and it's easy to see why. Because they are outside, the mess associated with wood heat stays outside. They don't require expensive chimneys, because the smoke that they produce doesn't need to be safely conducted out of your house. In the event of a chimney fire, the only peril to your house is probably from airborne burning ash.

Because outdoor boilers fall into the previously discussed furnace/boiler category, we've already covered most of their advantages, disadvantages and best applications, with 1½ exceptions: outdoor boilers are typically too powerful for a single residence, so their best application would be for heating a group of buildings. In that application, exception number ½ would probably be somewhat mitigated, since it refers to even greater inefficiency than that enjoyed by conventional woodburning furnaces. Exception number one might or might not be mitigated by the proper application: outdoor boilers are noteworthy for their smokiness, in part because they are so over-sized that they frequently run in low-fire mode, and, in part, because they are outside and jacketed with water (their usual heat-transfer medium) and, therefore, have difficulty reaching the temperatures required for efficient burning. The Canadian government tested the *overall* efficiency of a number of outdoor boilers and found many achieved less than 50%. Compare this to catalytic stoves, which test as high as 75%-80%. Some boiler manufacturers claim 90% *combustion* efficiency, which sounds impressive until you learn that all it means is that, if you load the boiler with 100 lb. of wood, it will eventually burn all but 10 lb. of ashes.

Fireplaces

The National Chimney Sweep Guild classifies fireplaces as "entertainment-oriented appliances," and this is a generally fair assessment. Fireplaces often remove more heat from the living space than they provide, because their large flues and un-  restricted access to combustion air allow most of the heat they produce to go up the chimney. Since in many homes the air the fireplace uses for combustion was heated by the furnace, the fireplace represents a net heat loss. (Somebody once suggested, only half in jest, that for home heating, most people would be better off playing a tape of a crackling fire in their VCR than lighting a fire in their fireplace.)

How, then, did our ancestors manage to heat their homes with fireplaces? There are several answers, the most important of which is that our ancestors' homes weren't very warm. A second answer is that their fireplaces probably did produce a net heat gain, because they weren't sucking furnace-heated air up the chimney or using it for combustion. Knowing this wouldn't have made you any warmer if you'd been trying to read by candlelight in the back bedroom of one of those colonial houses on a February night, but it does point to several improvements you can make in your conventional fireplace.

By installing glass doors and closing them when a fire is burning as well as when it's not, you sharply reduce the fireplace's demand for combustion air as well as its tendency to siphon warm air out of the house even when it isn't in use. (Most dampers are far from airtight, and none are insulated.) A second important step that you should consider if you have or are contemplating having a fireplace is to provide cold air returns which are built-in channels extending from near the fireplace (often the inside edge of the hearth) to the outside. These allow the fireplace to draw its combustion air from outside; the air there is, one would hope, more plentiful than it is in your house and you haven't, one would also hope, paid to heat it. We'll discuss cold, or outside air returns in a bit more detail in Chapter 7.

An option you might consider, if you haven't already got a fireplace and can't bear to be without one any longer, is to have yours built to the 200-year-old specifications of Count Rumford. Rumford, born Benjamin Thompson, was an American Tory who at the time of the Revolution emigrated to England, where he devised ways to improve the performance of the typically smoky fireplaces of the day. The Rumford fireplace, long neglected but making a comeback, differs from a typical low-linteled, deep fireplace in that its high, wide and very shallow firebox and low fireback effectively radiate its heat into the room. The placement of the chimney flue and the design of the throat (damper opening) assure a draft strong enough that, even with the fire built where it should be—partially on the hearth—the Rumford won't fill the room with smoke. While not nearly as efficient a heater as a stove or furnace, a Rumford supplied with outside air may provide a slight net heat gain while allowing you to see and hear the fire. You can have your cake and eat a little of it, too.

It is important to note that a true Rumford has very exacting specifications which must be followed if the fireplace is to live up to its reputation. Some "Rumford-style" fireplaces—look-alikes that alter the design proportions—are probably not any more efficient than conventional fireplaces. Also note that, because of their shallow fireboxes, Rumfords and many of their inexact copies are probably unsuited for glass doors, since there is potentially too much heat near the front of the firebox.

BEST APPLICATIONS

You aren't going to heat your whole house with one, so a fireplace is most appropriate for people to whom the esthetic value of fire is paramount, and the heat it produces is of little or no importance.

ADVANTAGES

The senses. Fireplace stoves and woodstoves with viewing windows allow a substantially smaller view of the fire than almost any open fireplace will provide. Further, the more efficient heaters don't allow you to hear or smell the fire, while the traditional open hearth most certainly does. Touch and taste are a wash, so we have a clear winner.

Little creosote. Fireplaces are usually clean burning, at least in comparison to most old airtight woodstoves, but it is also true that some of the steps that you can take to make a fireplace marginally more efficient, such as installing glass doors and regulating the burn rate with the damper, will produce more creosote.

DISADVANTAGES

Inefficiency. What else is there to say?

Price. This will vary a good deal, depending upon what sort of fireplace you want, where you live, how tall your chimney must be, etc., but a good, mason-built Rumford will cost much more than a woodstove and will never amortize itself in any tangible way except, possibly, by increasing the resale value of your house.

Maintenance. This does not apply so much to masonry fireplaces, (though they, too, need repair from time to time), as to factory-built fireplaces with metal fireboxes. The metal used in these units is fairly thin sheet steel, prone to rusting, warping, buckling and damper problems, any of which can be difficult and/or costly to repair.

Masonry Heaters

There are a number of different kinds of masonry heaters: Russian fireplace, Finnish contraflow and tiled heaters from Germany and Austria among them—which, despite differences in design and appearance, exemplify an approach to wood heating that differs sharply from that of stoves and furnaces. Instead of a round-the-clock fire maintained by periodic stoking and control of the air supply—the modus operandi applied to other serious wood heating equipment—masonry heaters rely on very hot fires—at times in excess of 2,000 degrees Fahrenheit—of short duration. Fires lasting only an hour or two heat a masonry mass weighing a ton or much more. The mass then radiates the stored heat for 12 to 24 hours, depending upon the weather. The extremely hot fires result in very clean burns.

Masonry heaters reportedly surpass conventional heaters in their heat-transfer efficiency ratings—how much of the produced heat ends up in your house instead of going up the chimney. A conventional stove operating at high combustion efficiency may send up to half its heat up the chimney. A well-designed masonry heater, on the other hand, stores and radiates something on the order of 80% of the heat it produces. It does this by directing the intensely hot gases through a series of channels in the masonry mass. By the time the exhaust reaches the top of the chimney, it is almost cool, having left its heat in the masonry. The smoke does not deposit creosote if the heater is properly operated, because the fire is so hot that the tars and organic compounds are consumed in the firebox.

The various kinds of masonry heaters differ mainly in their heat extraction systems. The Russian fireplace uses an especially complex chimney—built with horizontal passages and a top-mounted damper to hold in the heat when the fire is out—to trap and store the heat. The fire could be built in almost any fast-burning appliance—a kitchen or sauna stove, a fireplace, a water heater—attached to the specialized chimney. The Finnish contraflow, on the other hand, is a self-contained firebox and heat storage unit that is vented into a conventional chimney.

As intriguing as their efficiency makes them, masonry heaters are made even more intriguing, at least to me, by the vintage of the technology, which originated in ancient Rome and saw widespread application in northern Europe and Russia in the 1700s.

BEST APPLICATIONS

Because masonry heaters are large (a contraflow's "footprint" might typically be 3-by-5 feet) and very heavy, installing one in an existing house will often prove to be an expensive proposition, assuming that it's possible at all. A new home, however, can be easily designed around a masonry heater. Sized to your house, a contraflow or Russian fireplace should provide all the heat you need and, because of the gentle, even nature of the warmth (the surface temperature of a contraflow seldom exceeds 150 degrees Fahrenheit), will be more akin to central heating than space heating.

ADVANTAGES

Beauty. Masonry heaters, faced with stone, brick, tile or other masonry materials, provide a full measure of esthetic enjoyment. If you wish, you can have one built with a window so that you can see the fire.

Utility. A masonry heater is arguably the most effective and efficient wood heater you can buy.

Ethics. Environmentally, masonry heaters may outshine all other woodburners and, perhaps, every heating system save solar: they produce a very small quantity of emissions and will use a quarter to half the wood that a stove would require to heat the same space.

Low maintenance. Because masonry heaters are so clean burning, your chimney will stay cleaner than it would with most woodburning appliances. Also, there is little in the system that is likely to wear out.

DISADVANTAGES

Price. A basic manufactured contraflow kit will cost nearly twice as much as a top-of-the-line catalytic stove, and this does not include the chimney, the footing or floor support, the exterior finish, or the labor. Scandinavian imports will set you back well over $10,000. Over a long period of time (maybe 20 years), however, a masonry heater should amortize itself, even when compared to the most efficient woodstoves, with the high initial cost being offset by reduced fuel and chimney cleaning requirements and the elimination of the need to replace catalytic converters.

Availability. Clearly, you can't go to your local stove store, pick out the model of masonry stove you want, and bring it home in the trunk of your car. As of this writing there are masonry heater kits available from a Canadian firm, Temp-Cast, of Port Colborne, Ontario and possibly from others. There are also a few masons scattered hither and yon who specialize in masonry heater construction. Make sure you hire one who knows what he or she is doing; an incorrectly designed or constructed heater is costly and useless at best, and costly and dangerous at worst.

Convenience. If you bridle at the thought of kindling one or two fires from scratch every day, a masonry heater will seem inconvenient. Also, a totally cold heater may take hours after it has been fired to produce significant heat.

Esthetics. A masonry heater can be, and usually is, a thing of beauty, but used as it should be for maximum efficiency, it won't provide you with the sight of a cheery fire most of the time.

Quest Plus pellet stove from
Whitfield Hearth Products

Pellet Stoves

As concern over the environmental impact of woodburning has increased, so too has the stove industry's attention to mitigating that impact. Much attention has been directed at designing new stoves, but some has also been given to designing a new fuel: the pellet. Pellets, which look like nothing so much as rabbit food, are sometimes made of waste wood but, depending upon the availability of raw materials, may be made from agricultural waste or cardboard. Pellets require special stoves equipped with electric-powered forced air intake, and this, combined with the fuel's low (5%) moisture content, leads to very low emissions and to overall efficiency ratings that are claimed to be a bit higher than those of catalytic stoves. Because the draft is forced and the exhaust gases are relatively cool, it is possible, depending upon the manufacturer's recommendations and local building codes, to vent a pellet stove with a relatively inexpensive double-walled pipe in lieu of a chimney.

In recent times more than half of the stoves being sold in the northwestern United States were pellet stoves, and their popularity is increasing in Canada, as well.

BEST APPLICATIONS

People who live in areas unsuited by climate and/or topography to conventional woodburning equipment may find pellet stoves appropriate, as will people without ready access to firewood. Pellet stoves will also appeal to those seeking convenience: the operator must reload the hopper from time to time, but the stoves are otherwise self-stoking.

ADVANTAGES

Ethics. Pellet stoves produce little in the way of emissions and burn fuel that would otherwise probably be waste.
Looks. Models with glass doors allow you to see the glow of the flames.
Cheap and simple installation. A stove that doesn't require a conventional chimney can save you a bundle.

DISADVANTAGES

Price. Pellet stoves usually cost more than woodstoves, and the fuel isn't cheap. As a national average, pellets currently cost about $3 per 40 pounds, or $150 per ton. With 1 ton of pellets having the heat value of 1½ cord of hardwood, fuelwood must cost $100 per cord in your area for pellets to be an economical fuel.
Dependence on electricity. When the power goes off, so do most pellet stoves (a few have battery back-up), since they lose their feeding auger and forced draft. I can live with candles and flashlights, un-

washed dishes and unflushed toilets, but I insist upon being warm. If you live in an area subject to power outages, you might want to think twice before you buy a pellet stove.

Availability of fuel. If you live in an area with a sufficiency of pellet mills, no problem. In my area, we had one, but it shut down, leaving the people who bought stoves with the choice of paying a premium for fuel trucked in from afar, or abandoning their expensive purchase.

Stove Installation

I would guess that it's a rare chimney sweep who doesn't encounter a large number of stoves that are, for one reason or another, unsafely installed. My own very unscientific survey indicates that at least 50% of the installations I see for the first time pose significant hazards to the people who share the house with them. Solving these problems is sometimes as simple as securing the joints of a stovepipe with sheet metal screws, but often it is far more difficult. Old houses with old chimneys and stoves are frequent offenders, but new houses are far from immune.

In most cases, unsafe stove installations are dangerous because the appliances are located too close to combustible materials. The kindling temperatures of walls, ceilings, floors and furniture that are too close to stoves and stovepipes are reduced over time. That means they can ignite more easily than objects located a safe distance from the stove. It's true that these hazards are theoretical, but it's also true that they sometimes realize their destructive potential outside the testing lab. The odds of your not-to-code system's setting your house on fire may not be great, but the consequences if it did could be catastrophic.

As a consumer, you should be able to depend upon a certified chimney sweep to recognize whether or not your stove is safely installed (though you should bear in mind that some unsafe conditions are hidden from view), and, perhaps, a knowledgeable insurance inspector will know what to look for. But a surprising number of people who should know the safety requirements—masons, building contractors, firefighters—often do not. If you have a stove or are contemplating having one installed, it therefore behooves you to know enough about safe stove installations to avoid dangerous mistakes and recognize when to get help.

Note that we're assuming your chimney is suitable; it may not be. We'll discuss this in the next chapter. Note also that all new stoves have, or should have, specific instructions in the owner's manual and on a tag attached to the stove concerning their recommended clearances and installation requirements. These instructions supersede anything that I say here, as do local and/or state or Provincial building codes.

CLEARANCES

Stoves. Unless the manufacturer's instructions say otherwise, a stove should be 36 inches (48 inches by Canadian codes) from anything combustible (except for the floor—more on that shortly). This includes walls, drapes, chairs, ceilings, mantelpieces—anything that can burn. By installing a heat shield—a sheet of noncombustible material such as sheet metal or masonry, spaced 1 inch from the combustible surface and open on the top, bottom and sides to allow free air circulation—the 36-inch clearance can be reduced to 12 inches from walls and 18 inches from ceilings. Take care, however, that the shield is large enough to cover any area within 36 inches of the stove. Some stoves have built-in heat shields which reduce the clearance necessary behind the stove. Install these stoves (and all others) according to the manufacturer's specifications; the built-in heat shield probably doesn't obviate the need for additional shielding in some situations.

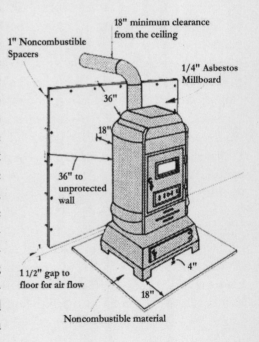

18" minimum clearance from the ceiling

1" Noncombustible Spacers

1/4" Asbestos Millboard

36"

18"

36" to unprotected wall

1 1/2" gap to floor for air flow

4"

18"

Noncombustible material

Stovepipe. Single-wall stovepipe should be 18 inches from any combustible surface. By installing a heat shield—either on the pipe or on the surface being protected—this distance can be reduced to 6 inches from a wall and 9 inches from a ceiling. As with the stove heat shield, a pipe heat shield should have a 1-inch ventilated air space for maximum clearance reduction. There is double-wall stovepipe available that is approved for 6-inch clearance with 6-inch pipe (8-inch clearance for 8-inch pipe) . Be careful that the reduced pipe clearance made possible by shields or double-wall pipe doesn't bring your stove itself too close to anything combustible.

Understove Protection. Again, new stoves will be labeled with their clearance requirements, but in the absence of these, the protection, known as the stove board, should extend 18 inches beyond the stove in all directions. Its composition should be as follows: For stoves with legs 6 inches long or longer, the understove protection should be 2-inch-thick masonry (bricks, usually) covered with 24-gauge sheet metal. For stoves with legs 2-6 inches long, the masonry should be 4 inches thick, and hollow, and laid so that the hollows form continuous open-ended passages for air circulation. The sheet metal covering is the same as for the other board. Stoves with legs shorter than 2 inches must not be installed on combustible floors at all. Most stove dealers have or can get manufactured stove boards in various sizes, and half-inch Homosote NCFR, or fire-resistant, wallboard covered with sheet metal is understood to provide adequate understove protection, though it is not specifically intended for this purpose.

Many otherwise careful people dismiss the need for special floor protection, saying something like, "Oh, my cat sleeps under the stove all the time. How hot can it be?" I would point out that the cat, unlike the floor, is mobile (though I've known many tom cats who were barely so), and can get away from the stove and cool off. Also, a cat wouldn't begin to smolder without letting you know about it; a floor would.

INSTALLING STOVEPIPE

Besides the clearances, there are other things worth knowing about stovepipe:

Less is more. Pipe releases a surprising amount of heat, in many installations as much as the stove itself. A large amount of pipe, therefore, encourages creosote formation by allowing the smoke in the pipe to cool before it reaches the chimney and by encouraging you to run the stove slowly, since you get so much heat from the pipe. The National Chimney Sweep Guild (NCSG) recommends not using more than 8 feet of pipe in an installation, unless it is unavoidable, in which case, the pipe should be supported every 4 feet with a solid bracket or hanger.

Concerning 90-degree pipe elbows, less is more, also. Smoke is like a car—corners slow it down, encouraging cooling and creosote deposition, so use no more than two elbows, not including the chimney's thimble or the stove's collar.

Stovepipe comes in different gauges. The lower the number, the thicker the pipe. The NCSG recommends 24-gauge pipe (.023 inches thick) for most stoves, and it certainly won't hurt to use 22-gauge (.029 inches). Pipe of these thicknesses stands up to daily use and to pipe fires much more reliably than does pipe in the lighter gauges 26 (.019 inch) and 28, and will usually last enough longer to offset its higher price. Another good cautionary note from the Guild: don't use galvanized stovepipe in a woodstove installation, because it gives off zinc vapor, which is toxic, at temperatures of 750 degrees Fahrenheit and higher.

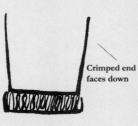
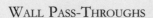
Crimped end faces down

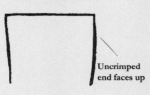
Uncrimped end faces up

Horizontal sections of pipe should, ideally, slope up toward the chimney at a rate of $1/4$-inch for every foot of length. They should never slope down. Also, the seams should be on the top side of horizontal sections to prevent liquid creosote from dripping out.

Vertical pipe should be installed with male (crimped) ends pointing down, again to prevent creosote dripping, which is not just unsightly, but is a potential fire hazard, as well.

Joints. Each pipe joint should be secured with at least three sheet metal screws for stability, particularly in the event of a red hot shaking pipe fire. Burning stove pipe falling suddenly on the floor calls for quick, correct action; better to not test yourself.

WALL PASS-THROUGHS

Connecting a stove to a chimney through a combustible wall is tricky business. It's easy to be fooled by noncombustible facades such as brick or tile hearth walls, but you need to look closely, bacause if your chimney is on one side of a combustible wall and your stove is on the other, you likely have a problem. Very seldom, even in new houses, do I see the problem safely dealt with. As of this writing, the National Fire Protection Association recognizes only four ways of connecting a stove to a chimney through a combustible wall:

A *clay tile thimble* surrounded by 1 foot of solid masonry on all sides is safe. You rarely see this—though thimbles with much less than 1 foot of surrounding masonry are commonly used—because such a pass-through is large: approximately 32 inches square if a thimble with a 6-inch inside diameter is used.

A *section of factory-built, mass-insulated metal chimney* with a 2-inch air space surrounding it and a single-wall stovepipe installed through the center of the chimney section, with a 1-inch air space surrounding it (an 8-inch chimney section with a 6-inch pipe, for example) can be used, though centering the pipe is difficult.

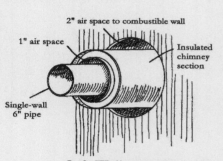

2" air space to combustible wall

1" air space

Insulated chimney section

Single-wall 6" pipe

Safe Wall pass-through

The same section of factory-built chimney with a 9-inch air space all around it may be safely used without the smaller pipe passing through it. Confusion is common because factory-built chimney is typically approved for 2-inch clearance, but this clearance is safe when this product is part of a vertical chimney, or is part of an engineered wall pass-through kit for a chimney of the same material—not a masonry chimney. Note also that a chimney section used in this way should not simply be butted up against the masonry chimney; it should make a positive, tight connection with it (though it shouldn't actually project into the flue).

A *ventilated metal thimble* with two (not one) 1-inch air passages completely surrounded by 6 inches of noncombustible insulation is acceptable, but since, as of this writing, no such device is commercially available, this option should be disregarded.

I have seen a heavily insulated thimble which was tested and listed by Underwriters Laboratories for installation with a 2-inch clearance; an excellent solution to an often vexing problem, but, alas, the last I heard it was at least temporarily out of production.

FIREPLACE INSTALLATIONS

Many houses built after central heating became commonplace have fireplaces, but it wasn't until after the oil crisis of the early 1970s that woodstove hookups to fireplaces became commonplace. When the oil crunch struck, many people bought woodstoves and, in one way or another, installed them in their fireplaces. There were two commonly used techniques for doing this:

1. Stovepipe was connected to the stove and extended just past the fireplace damper, and the remaining space in the damper opening was filled with fiberglass insulation or a sheet metal plate.

2. A metal plate that covered the fireplace opening was installed and the stovepipe passed through a hole in its center. Several manufactured stoves had this plate built in.

The problem with both installations is that smoke from the stove doesn't go directly into the chimney flue but lingers and cools in either the fireplace or its smoke chamber. When coupled with fireplace flues that are usually too large for a woodstove, the result is often a large buildup of creosote, particularly glazed creosote, in the chimney and fireplace. I vividly recall removing over 40 gallons of creosote from such an installation some years ago.

A large fireplace flue should be lined to downsize it for a woodstove, and it should make a direct connection to the stove, but many people resist doing this, either because of the expense involved, or because the smaller flue makes it impossible to use the fireplace as a fireplace. There are ways of installing a stove in a fireplace which, while not addressing the problem of the flue's being too large, do result in smoke's entering the flue without cooling in the fireplace or smoke chamber and, therefore, reduce the risk of dangerous creosote buildup. Before discussing these methods, I want to stress that, in my opinion, the only good way to install a stove or insert in a fireplace is by relining the flue and directly connecting the stove to the new liner. Codes in Canada now wisely require this, but this is not generally the case in the U.S. For readers below the border who insist upon ignoring my advice, here are three ways of installing a stove in a fireplace without relining:

1. *Use a separate thimble.* Often the cheapest and, in my opinion, often the best method is to install a thimble that enters the chimney flue, usually over the mantel. The damper is then shut and sealed with high-temperature silicone, or replaced if the damper is too warped to effect an airtight seal. The main drawback to this type of installation is that wooden mantels, trim and paneling often present difficult clearance problems. A second drawback is that the installation is semipermanent:

Seal space in damper opening

the fireplace should not be used unless the thimble is filled with masonry and mortar.

2. *Use a connector kit.* These are usually flexible stainless steel oval-shaped pipe which can be pushed up through the damper opening to join the chimney flue. The unfilled space in the damper is filled, usually with sheet metal. The oval pipe has a round thimble to accept the stovepipe. The main drawbacks of this method are cost—the kits cost several hundred dollars— and maintenance. It can be a real pain to remove and reinstall these kits, but it must be done at every cleaning.

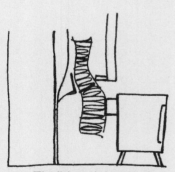

Flexible stainless steel installation kit

3. *Use more pipe.* The third technique is simply to extend stovepipe through the damper opening all the way to the beginning of the flue. The unfilled space in the damper opening is sealed with sheet metal and high temperature silicone. The main problem with this method, aside from difficult maintenance, is that few fireplaces will permit it, either because the damper opening is too small for round pipe, or because the line from the stove to the flue is too indirect, or both.

Note that just clearing the damper opening with the pipe isn't good enough; the pipe must reach to the first flue tile, or else too much creosote will build up and too much heat will radiate to surrounding wood through the unlined walls of the smoke chamber.

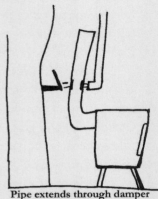

Pipe extends through damper opening to first flue tile

One important safety note concerning stove-in-fireplace installations: manufactured metal fireplace and chimney units—known as "zero-clearance" fireplaces—are engineered and approved for use as open fireplaces, but cannot accept woodstove or insert installations or, for that matter, aftermarket glass doors. Such modifications could cause excessive heat buildup which could transfer to nearby combustible walls. If you have a manufactured fireplace, use it only as detailed in the manufacturer's instructions. If you are unsure whether your fireplace is a "zero-clearance" type or not, have a certified chimney sweep check it out.

Stove Operation Tips

Let me begin this discussion of how to operate woodstoves, furnaces and fireplaces with a large disclaimer: nothing that I say is intended to supersede the instructions and advice found in your appliance's owner's manual, or the advice you receive from a knowledgeable stove dealer or chimney sweep. In short, the advice that I offer is purely generic, as it must be, since I have no specific experience with your particular heating system.

I speak primarily to woodstove owners, with whom I group those who have wood furnaces; closed fireboxes and limited oxygen supplies have a greater potential to contribute to operational problems than do open fireplaces.

After years of burning wood myself and observing and dealing with other people doing the same, I have come to think of the pursuit as being akin to tightrope walking, balance being the element crucial to the success of either endeavor. The balance I'm speaking of is between the cleanest possible burn and the burn that provides you with the greatest amount of usable heat and therefore uses the smallest possible amount of fuel. Absent the technological assistance of catalytic converters and effective secondary combustion, the tightrope is a stark delineation: slip off on one side and your chimney stays clean, your fuel consumption increases, and your stove will neither hold a fire long enough for convenience's sake, nor in all likelihood will it heat your living space without constant refueling and overheating. Slip off on the other side of the tightrope, and you have long-lasting fires that contribute much of their heat to your home, but are smoky and lead to a dangerously dirty stovepipe and chimney.

For people with catalytic or high-tech stoves, the tightrope becomes less distinct but doesn't disappear. These stoves, it is true, consume part of the smoke produced by the fire, but miraculous as they seem, they still require oxygen for combustion: too little will result in inefficient heating. A catalytic or high-tech stove may make the best of a bad situation, but operated too far from the tightrope—on either side— neither will perform as it should.

For the purposes of this discussion, let's assume that your stove, house and chimney are well matched to one another, the weather is cold enough to make woodburning sensible, and your wood is properly seasoned. This leaves you, the operator, and the critical variables that govern combustion and over which you have immediate control: heat, oxygen supply and fuel. Complete or, at least, good combustion occurs in three stages:

1. *Drying,* during which enough moisture is cooked out of the wood to allow it to burn;

2. *Pyrolysis,* during which the wood's molecules are changed by exposure to heat, producing tar droplets and gases;

3. *The charcoal stage,* which is the combustion that occurs after the tars and gases have either burned or exited the firebox.

The three stages can and often do take place simultaneously in different parts of your stove, or even in different places on the same piece of wood. All three of the stages, indeed the entire combustion process, require heat and oxygen; deprive a fire of either at any point in the combustion process, and it will soon go out.

HEAT

Strike a match, and you have heat, but not enough heat to drive the moisture from a firebox full of wood. This is why you start a fire with paper and kindling and gradually add larger pieces of wood only after the kindling and succeeding sticks are burning well: wood in the first stage of combustion doesn't produce enough heat to drive the moisture from additional wood added prematurely to the fire. By not overloading the firebox, you assure that there is always wood in the heat-producing second and third stages of combustion.

OXYGEN

Throughout combustion, the temperature of the fire is determined in part by how much oxygen is available to it. A fresh load of wood in the stove requires, for a short time, as nearly unlimited a supply of oxygen as the stove's design allows in order to evaporate the moisture in the wood and proceed to the next stage of combustion, at which point you are again on the tightrope. The fuel in the pyrolysis stage produces heat. At this stage, continuing to supply oxygen in unlimited or large amounts will, at least, make you uncomfortably warm for a little while. It will also rapidly burn up the load of wood, and, at worst, result in serious overheating and damage to your stove and/or chimney. Restricting the oxygen supply too much, on the other hand, will prevent the gases and tars produced in this stage from burning, which, of course, will result in creosote formation.

How can you tell if you are adding enough oxygen to the fire? First, by installing a stovepipe thermometer and learning to understand it: the surface temperature it measures is a bit more than half the flue gas temperature. Flue gas temperatures below 400 degrees Fahrenheit indicate inadequate oxygen (assuming that there is sufficient fuel in the firebox for a fire), and flue gas that is hotter than 900 degrees Fahrenheit is potentially damaging to your chimney. Your stovepipe thermometer, then, should read between 250 and 475 degrees through most of the burn cycle.

A second good indication of how well managed your fire is will be the smoke coming out of the chimney once the fire is well established. Fires that have just been kindled usually produce a good deal of smoke for a short time. Thick gray, brown or black smoke is evidence of an oxygen-starved fire; light gray, almost white smoke with a good degree of transparency indicates that you are supplying enough oxygen for efficient combustion. It is often almost impossible to detect smoke coming out of the chimney of a well-run catalytic or high-tech stove except in exceedingly frigid weather when a small amount of condensing water vapor will be visible. It was, in fact, my amazement at seeing one of these "smokeless" chimneys that convinced me to get a catalytic stove.

A third way of judging whether your fire is getting the right amount of oxygen is to look at it. The fire should be active, not smoldering (which indicates too little oxygen) or roaring (which indicates too much). Note that, if your stove doesn't have a viewing window, it may be difficult to assess the fire this way, since opening the door gives it a sudden supply of oxygen it doesn't have during normal operation. It's still worth doing, though, because there are a couple of telltale signs of poor combustion that a sudden whiff of air won't disguise: the smell of creosote (rather than woodsmoke) in the firebox, and the formation of creosote inside the firebox.

Fuel

Fuel—in this case, wood—is the third ingredient necessary for combustion. We've already discussed what kinds of wood make good fuel, and will later discuss how to season and store wood, so what's left? How to put it in the stove, is the answer, and I don't mean just opening the door and tossing it in. How you manage the stoking of your stove will have a large impact upon how effectively it heats and how clean your chimney stays. Consider the opposite extremes of stove stoking techniques, the rationales behind them and some of their possible repercussions.

Woodburner A (not her real name) kindles a fire, fills the firebox as rapidly as possible, then lets the charge of wood burn entirely, or at least well into the charcoal stage, before repeating the process. She feels that she gets the cleanest possible burn this way because, 1.) She doesn't frequently open the stove door, thus introducing large amounts of room-temperature air and cooling the firebox; 2.) She isn't constantly adding room-temperature—or colder—wood, which will temporarily cool the fire; and 3.) She allows the fire to remain at its cleanest-burning stage—the charcoal stage—for a significant period of time. What she overlooks is that her chimney, particularly if it's exterior, may have a chance to cool between fuelings, increasing the possibility of creosote buildup. She also ignores the possibility that since she only adds wood when the fire is in a relatively inactive stage, she may unnecessarily prolong the first stage of combustion and encourage cool, smoky fires.

Woodburner B only needs enough kindling each year for one fire—his first, because from then on he adds wood to the firebox while there are still active flames. B's rationale is that by maintaining a nearly constant, active fire, he'll keep his chimney warm enough so that it will be unreceptive to creosote deposits. Also, by adding wood to an active fire, B feels that he encourages rapid evaporation of the moisture in the wood and prevents wide temperature swings. What he overlooks is that, 1.) Because he maintains a large load of fuel in his stove, he may have to overly limit its oxygen supply in order to avoid overheating; 2.) The overly full firebox that could result from his stoking procedure may inhibit the turbulence necessary to mix oxygen with the tars and gases given off by the wood during pyrolysis, resulting in incomplete combustion; and 3.) To a certain extent, the more wood you put in a stove, the more it will burn and the less completely it will burn it, perhaps because the entire fire is never given a chance to "burn down" in the charcoal stage. If you've ever increased your feeding of a stove to counteract a cold snap, you may have seen the evidence of this in the large increase in the quantity of ashes and unburned coals produced by the stove.

The stoking procedure that offers the best balance between the extremes and has the best chance of promoting a clean burn without excessive wood use or inconvenience is as follows:

Stoke moderately. After kindling the fire, never add more wood than the fire can handle. Putting a large, unsplit piece of wood on a small pile of kindling will result in a smoldering burn. Better to add small to medium-sized pieces of wood gradually to maintain an active fire. Don't be too stingy, though: a fire is most active when there is fuel layered above it, rather than laid in one row on the floor of the stove. As Daryle Thomas, an expert on solid-fuel heating (and not related to me), says, "When placing wood in a stove, think of a successful marriage: keep the logs close enough together to feel the heat and far enough away to breathe."

Use the stove's air controls. When the firebox contains sufficient wood (half to two-thirds full is a good place to stop unless you're stoking it for an overnight burn, in which case you'll probably have to nearly fill it), adjust the air inlets to produce the desired temperature, flame and smoke conditions described previously.

Reload, starting with fairly small, split wood, when most of the previous charge has burned down to large, glowing coals. Before adding wood, rake the coals toward the air inlet (often on the loading door) so that they form a smooth, consistent bed covering the bottom of the stove.

For a burn of relatively long duration, add wood as described—a few pieces at a time—until the firebox is full. Don't immediately close the air inlets; do it a bit at a time, and never close them all the way. You'll have to experiment to find the right setting for the air inlets, but recognize that your goal is not to have unburned wood eight hours after the final stoking; this would indicate that, for much of the burn cycle, there was too little oxygen present. Instead, be satisfied with a bed of coals adequate to start a new fire. Even taking the care described, a long-duration burn will probably produce creosote, but by only adding wood to an active fire and by not shutting the stove down until all of the wood is actively involved in flames, you ought to minimize the buildup. Many people try to atone for slow, smoky overnight burns by "burning the stove off" in the morning, *i.e.,* running the stove at a very high temperature for 15 or 20 minutes. This method is not nearly as effective as giving the stove adequate amounts of oxygen whenever it's burning, and is likely to start a chimney fire.

CATALYTIC STOVES

The above rules of operation are, for the most part, applicable to catalytic stoves as well as to high-tech, airtights and nonairtight stoves. The only significant difference is that, with a catalytic stove, you need to pay attention to properly preheating the catalyst before engaging it, since smoke passing through it will not ignite at temperatures lower than about 500 degrees Fahrenheit. This means that the temperature of the catalyst must be maintained at between 500-700 for 20 minutes after starting a fire. Most catalytic stoves have special probe thermometers that give accurate readings of the temperature in the catalyst. Follow the same preheating procedure for about ten minutes after refueling if the temperature has dropped below 500. Absent a probe thermometer, it's fairly easy to extrapolate from a stovepipe, or surface, thermometer. Locate it on the stove as near to the converter as possible, and remember to nearly double the reading—*i.e.,* a surface temperature of 300 probably means that the temperature at the catalyst is around 500-550.

HIGH-TECH STOVES

High-tech stoves also require preheating to effect secondary combustion. Consult your owner's manual for specific instructions.

FIREPLACES

As mentioned earlier, fireplaces are not as complicated to use as a stove can be, and unless yours has a problem related to design or construction, you either know how to use it already or will learn soon enough without my help. I will, however, describe four practices that I've found helpful in starting and maintaining fires in fireplaces.

Get the andirons out of the way. They may be decorative, but building a fire on them often results in logs smoldering above a distant bed of coals once the kindling is burned up. Keep the fireplace grate, though, and plenty of ashes to protect the bottom of the firebox from the intense heat.

To lay the fire, put a good-sized base log (3 or 4 inches in diameter) toward the back of the firebox, crumpled newspaper balls in front of the log, and a log cabin of cross-hatched kindling over the paper. When adding wood to the fire, add it so that it's cross-hatched with the ends of alternate rows of logs resting on the base log. This promotes good air circulation while allowing enough wood to stay in contact with the coals to maintain flames. And flames are what a fireplace is all about, aren't they?

Preheat a flue with a slightly recalcitrant draft (one that tends to smoke a bit during a fire's early stages) by holding a flaming newspaper torch above the damper just before you light the fire. Actually, you aren't so much preheating the flue as you are getting warm air moving up and stimulating a draft. Don't burn your fingers. A really balky fireplace may require you to open a nearby window slightly, since the balkiness may be due to an insufficient air supply in the house.

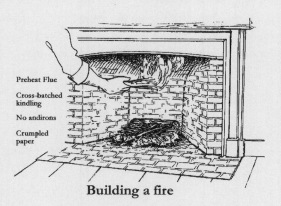

Preheat Flue

Cross-hatched
kindling

No andirons

Crumpled
paper

Building a fire

Use the damper. If your damper is adjustable, experiment with closing it part way as the fire burns low. If the draft is sufficient to keep smoke going up the chimney, you'll probably find that you get more heat in the room from the fireplace and burn a little less wood this way.

Ash Disposal

Wood ashes from a stove, furnace or fireplace may be cold after the fire has gone out, but—as, perhaps, with failed romance—they may also harbor glowing embers, and one such ember is enough. I know of a house burned to the ground by "cold" 3-day-old ashes shoveled into a cardboard box and left on the porch.

Always assume that ashes are hiding embers. Shovel them into a tightly covered *metal* container. Don't vacuum them up, and don't dump them where the wind can blow them onto your house or you neighbor's.

CHAPTER 5

Chimneys

If you burn wood, you need a chimney. This often neglected, frequently misunderstood and generally taken for granted (except by chimney sweeps) structure is, arguably, the most important component of any wood-fired heating system, with the probable exception of the operator. We're going to discuss chimneys now, but I cannot overly stress that a wood heating system is just that—a system. The same stove may not perform the same in two different chimneys; the same fireplace may smoke and smolder in one house but not in another; the same stove and chimney in the same house may behave differently for two different operators. And then there's the wood! It is essential, if woodburning is to provide the warmth and pleasure of which it is capable, that all the elements of the system—the chimney, the house, the appliance and the operator—work harmoniously together. If you are not currently a woodburner but are considering becoming one, you have quite a few choices and decisions ahead of you. If you already use wood, it won't hurt to consider your wood heating system. Perhaps your enjoyment of it could be enhanced, as well as its safety and efficiency.

New Construction

We'll begin by addressing those who are planning to build a new chimney for woodburning, whether as part of a new house or an existing house. As a chimney sweep, I frequently come across people who have had houses built with fireplaces or woodstove hookups that led to problems—major and minor—with their heating systems. The problems, I believe, stem from the process of the chimney's building. Typically, the only decisions the homeowner makes related to woodburning are along the lines of, "I want a woodstove in the living room," and later on, picking a color for the hearth brick. These decisions are rarely made based upon an understanding of their future implications, and neither, unfortunately, are subsequent decisions concerning such critical matters as the specific location of the chimney and its flue size, even though these decisions are usually made by architects or contractors. This process does not consider the wood heating system in any unified way and can result in houses with expensive and attractive but unusable wood heating components. This is a pity because, particularly in the case of a new house, good planning can make wood heat especially easy to live with.

I recently saw a case in which a family had a beautiful house built with a chimney in the living room for a woodstove. They moved in, bought and installed a stove and, when the weather got cold, fired it up. What they got, along with their heat, was a large amount of liquid creosote dripping on the hearth every time they started the stove and an alarming buildup of flaky and glazed creosote in the chimney. There were three—possibly four—reasons for the problem.

1. In the first place, the chimney was located on an exterior wall, thus assuring that it would rapidly cool the flue gases produced by the stove, resulting in the condensation and deposit of tars and creosote. Why was the chimney on the outside wall? Because that was where the architect's drawings placed it, probably to save space.

2. The flue was oversized—too large for a high-tech stove, which requires a chimney with draft velocity rather than volume. This was a decision made by the masonry contractor who, not knowing what kind of stove was going to be used, built a chimney that could accommodate—he thought—any stove.

3. The hearth design and thimble placement were at fault. The stove was placed in a mock fireplace with the thimble directly above it, facing down on the stove. While it's true that this arrangement eliminated the need for a stovepipe elbow, which is good, it's also true that it eliminated the possibility of a positive, dripless connection between the stovepipe and the thimble, with the result that a fair portion of the aforementioned condensing creosote and tar dripped onto the stove and hearth. Why was this design used? Because the masonry contractor had seen a picture of it in a trade periodical.

4. The fourth factor—and it may not prove to be a problem after all, though in different circumstances it would—was the stove itself, which may be too large for the very tightly constructed and well-insulated house. As suggested more than once above, if the stove makes the room too warm when it is operated properly, it will probably be damped down too far and too often for clean burning.

In this case the solution to the creosote problem was fairly simple, though expensive. The flue size was reduced by relining with an insulated stainless steel liner, and I'm happy to report that the wood heating system is working as it should. I'm not so confident about the other system—the architect-contractor-homeowner system—which caused this problem in the first place.

Almost any chimney sweep you ask will have had numerous similar experiences. The other common—and frustrating because they are so easily avoidable—problems that I often encounter in new construction are chimneys that do not allow a stove and/or smokepipe to be installed with the proper clearances to combustible materials, and fireplaces that smoke because no provision was made to give them air.

You can avoid costly, infuriating and dangerous problems with wood heating systems in new construction if you know enough to design the system as a whole. To a degree, the house needs to be designed around the system, rather than vice versa. I strongly suggest that you hire a qualified chimney sweep or other wood heating specialist as a short-term consultant. The money you pay him or her will be a drop in the bucket compared to what it would probably cost to fix a system that doesn't work properly.

In order to help you with the decisions you'll need to make when you're choosing a wood heating system for a new home, here are some general bits of advice:

Keep the chimney inside. Chimneys that are in the house for most of their length stay cleaner, work better, last longer and return more heat to the house than do chimneys outside the exterior walls.

Stove size is important. Woodstoves should be sized to the space they must heat. A stove too large for its space will provoke you to burn it slowly, causing creosote buildup. A stove that is too small may lead you to overfire it, possibly damaging the stove, the chimney and, should either fail, the whole house.

.785 x B² = Area

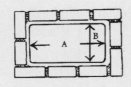

.785 x B² + (A-B) x B = Area

So is flue size. The chimney flue should be sized to the appliance it serves. Ideally, if your stove has a 6-inch round outlet, your chimney's flue will also be 6 inches in diameter. A larger flue may work well, but at a certain point (depending upon the height of the chimney, the stove itself, and other factors specific to each situation), a flue that is much too large for a stove will result in problems such as sluggish performance and rapid creosote buildup. A flue too small for a stove, though unusual, is unacceptable also, and will cause the stove to smoke. Fireplaces do sometimes have undersized flues. Since it relies upon a large volume of air to maintain its draft, a fireplace with an undersized flue will admit more air than the chimney can expel, with the result that smoke will back up into the house.

The area of a square or rectangular chimney flue is not as easy to calculate as a person with my primitive math skills might assume. A nominal 12-by-8-inch flue does not have an area of 96 square inches, in the first place because its inside dimensions are probably only 10½ by 6½ inches and, in the second place, because the corners of the flue are rounded. The flue's actual area is a good deal closer to 70 square inches. The diagram above gives the formulas for determining the areas of square and rectangular flues. A round flue's area is determined by multiplying its radius squared by 3.14, or *pi*.

Any competent mason will know the proper size relationship between the flue and the fireplace opening. (If you are interested, the area of the flue should be approximately one-tenth that of the fireplace opening. A short chimney may require a somewhat larger flue.)

Consider providing additional air. Fireplaces and, rarely, stoves may need cold air returns to work properly in tightly constructed houses.

Chimneys should be accessible for cleaning from both the top and the base. If access is difficult, it may be only your chimney sweep's problem (though it could cost you more). If access is impossible, it's your headache.

Block chimneys. Solid concrete blocks are an acceptable chimney construction material, but shouldn't be used as the exterior layer where they'll be exposed to the elements, because they can deteriorate rapidly, more than offsetting their low initial cost.

Each woodburning appliance should have a separate flue. This is a provision of the National Fire Protection Association code, but is widely disregarded, since multiflue chimneys are much more expensive than single flue chimneys. The risks of sharing flues include inadequate draft and the possible venting of toxic gases into the living space, the latter because many flues are only large enough to adequately handle the exhaust of one appliance; therefore excess gases in the flue will likely not make it out of the house. Also, if both your wood heater and furnace use the same flue, a flue blockage will leave you without a heat source until the problem is rectified.

All clearances must be respected. Chimneys and thimbles (the opening in the chimney for the stovepipe) must be located so that the appliance can be installed in accordance with clearance requirements. It's shockingly common to find, even in new houses, thimbles 4 or 5 inches from a combustible ceiling; they should be at least 18 inches below it.

Minimize pipe. The less stovepipe an installation requires, the better. Smoke cools rapidly in the pipe, leading to creosote formation. Having less pipe will allow you to run the stove hotter since you'll get less heat from the pipe.

Factory-built, insulated stainless steel chimneys approved for use with wood heating appliances are a perfectly reasonable choice. Because they are well insulated and well sized to modern stoves, they tend to stay cleaner, in my experience, than masonry chimneys. A word of caution, though: Stainless steel, when exposed to intense heat such as that associated with a chimney fire, undergoes molecular changes and loses its ability to withstand the heat from subsequent chimney fires. Masonry chimneys are frequently damaged or destroyed by chimney fires, as well, but the damage there is visible. It won't necessarily be with a metal chimney, so the only prudent course of action is to replace it. This is true even though these chimneys are tested to withstand temperatures of 2,000 degrees Fahrenheit or more: the test lasts about ten minutes; a chimney fire, or, more accurately, the heat it generates, rarely goes away so quickly.

Existing Chimneys of Recent Vintage

This section is for people with fairly new and apparently sound chimneys. It's quite a temptation, as we enter another age of high-cost energy, to plug a woodstove into your chimney and enjoy cheap heat again. Before you run out and buy a stove, though, you need to spend some time evaluating your chimney. Again, I suggest that you hire a qualified sweep to inspect it and make recommendations, not because I'm trying to drum up business, but because I've come to realize that a high percentage of existing chimneys, unless they are modified, are unsuitable for wood heat. A good sweep can save you from having a lot of serious problems. Here are some of the factors he or she will consider:

Liner. Since the chimney is fairly new, it's probably lined with clay tile, but is the lining sound? Are there cracks, loose mortared joints or other deterioration?

Soundness. If your chimney is factory-built metal, is it appropriate for the type of appliance you wish to vent with it? Is it sound (hard to tell, unfortunately, unless you know there have been no chimney fires in it)? Were required clearances adhered to in its construction? Were all of the necessary support and shielding components used?

Adequate flues. Will the chimney provide a flue solely for the use of the appliance you intend to install?

Clearances. Will the location of the chimney and thimble allow for a safe installation--one that abides by required clearance standards? If not, can the problem be rectified?

Right kind of stove. Are there inherent factors, such as flue size and chimney location, that may affect your choice of stoves?

Pass-throughs. If the chimney and stove are on opposite sides of a combustible wall, does the wall pass-through provide adequate clearance? Don't be lulled into a false sense of security by brick or stone facades; be certain that the connection is safe.

Fireplaces. If you are considering activating an unused fireplace, you need to evaluate the condition of the flue, damper, firebox and smoke chamber, as well as the chimney. Was construction done according to applicable codes? Is the system free of leaves, birds' nests and other debris?

Old Chimneys

You own a lovely old farmhouse in the country. The views and solitude are magnificent, the sills and beams are fairly sound—and you spent $3,000 on heating oil your first winter there. Come spring, you hop in your truck, drive to the nearest stove store and buy the latest, most efficient woodstove your dwindling cash reserves can afford. You bring it home along with a few sections of stovepipe and a fireproof hearth mat, call a couple of rugged friends, and install the stove in the handsome old brick kitchen chimney. Now all you need are a few cords of wood, and your new purchase can begin to amortize itself.

Then it occurs to you that the chimney probably is dirty, so you call a chimney sweep, who's glad to check your chimney. He arrives, sets up ladders and, instead of sweeping, gives you some bad news: your chimney isn't lined. It is one layer of bricks held together by crumbling hundred-year-old mortar, its own weight and inertia. The sweep explains that the house and chimney were built long before it was common practice to line each flue with clay tiles. Furthermore, your chimney can't be used safely to vent your stove until it is lined; excessive heat, sparks and even flames can escape from your chimney and kindle a fire in areas of your house other than the stove's firebox.

What's more, the chimney sweep continues, an unlined chimney can't be cleaned as well as a lined chimney; his brush won't reach the nooks and crannies between the bricks where creosote collects.

After the sweep leaves, you ponder the problem. Maybe he's being an alarmist: people burned wood for hundreds of years without lined chimneys. But old-timers didn't have airtight stoves. Airtights can cause a greater and more rapid buildup of creosote than is common with leaky old-fashioned stoves, and even clean burning catalytic stoves may perform poorly in old chimneys, since the flues are liable to be too large and the rough, unlined interior will provide too much resistance to the smoke. Also, you've noticed a few burned-out cellar holes here in wood heating country. You envision your house as a charred ruin and decide to have the chimney lined.

How can you tell on your own whether your chimney is lined? If the house was built prior to the 1940s, there's a good chance that the chimney is not lined. Take a look for yourself. If the chimney isn't too dirty, you'll probably be able to tell immediately whether it is lined. Look down the chimney with a good flashlight, or look up from the cleanout or thimble using a mirror (a bright, sunny day when the sun is high is the best time for this). If the chimney is unlined, the inside will look like a brick wall. If it's lined, you'll see smooth tiles, not bricks, and all of the joints will run horizontally. If you aren't sure what you're seeing, it's time to call a chimney sweep for an inspection.

The presence of tile liners does not automatically mean that the chimney is completely lined; some old chimneys have been rebuilt from the roof line up with tiles but remain unlined from the roof down through the house. Nor should you assume that a tile-lined chimney is in good enough repair to be used; time, chimney fires and the normal temperature swings associated with wood heating can cause the mortar between the liner's joints to deteriorate and fall out, leaving gaps in the protection. Water, earthquakes and normal settling can also render a chimney that appears sound unsafe. I tend to trust recently built tile-lined chimneys, but I've learned that with chimneys, as with most things, looks can be deceiving and trust fleeting. Thoroughly inspecting a chimney before using it and periodically thereafter is a sensible precaution to take.

On request, a chimney sweep can inspect a chimney after he cleans it by lowering a strong light through it, if the chimney is straight and not too tall, or by sealing the chimney and inserting a smoke bomb. Either method will probably reveal major damage if it's present, and the light may reveal some incipient problems, but more definitive inspection probably requires the use of a miniature video camera specially adapted for the job. This can be an expensive procedure and, given the cost of the equipment involved, many sweeps don't do it. In my experience, many of the sweeps who do video inspections also do a good deal of relining. This gives them an ax to grind and may detract from their credibility, inasmuch as every unsafe flue revealed is a potentially profitable job. If you are in doubt, don't hesitate to get a second opinion, but don't hesitate to get the video inspection in the first place; "totally definitive" and "chimney" seldom, if ever, belong in the same sentence, but a well informed guess is almost always worth having.

Lining Options

In an age that has produced such technological marvels as microwave ovens and inflatable basketball shoes, it may come as a surprise to some readers that only three materials are properly used for lining chimneys: thermal concrete, stainless steel and clay tile. Each method has advantages and disadvantages, and there are subgroups within the three major groups. Note that, with any of these methods, if the job is a relining—that is, the replacement of a damaged tile liner—then the first step will be the removal of the damaged tiles. This can be tricky, requiring specialized tools in many cases, and is probably beyond the scope of all but the most determined and resourceful do-it-yourselfers.

Thermal Concrete

The process of installing a concrete chimney liner usually—but not always—involves placing an inflatable bladder in the chimney and pouring a special concrete mixture into the chimney around it. Once the concrete has set, the bladder is deflated and removed and—presto— the chimney is lined. Most of the companies offering concrete lining use this method, but at least one—The Ahrens Company—fills the chimney with no-slump concrete (concrete with a very low moisture content that stays where it's poured) and then pulls a vibrating bell up through to form the flue.

The concrete used for chimney lining differs from the material used to make your basement in this respect: instead of mixing sand and gravel with water and cement, chimney lining companies use lightweight, granular insulating material such as perlite in place of the sand and gravel. The result is concrete with very good insulating quality (good for maintaining warm flue temperatures) that is relatively light and so is less likely to induce the collapse of an old chimney should the mortared joints between the bricks have lost their integrity (though it would be wise to address that problem.)

ADVANTAGES

Effectiveness. A warm flue tends to stay cleaner and draw better than a cold flue, and is less subject to damaging condensation. In addition, the liner is smooth and seamless, thus providing very little obstruc-

tion to smoke traveling through the chimney and offering no joints
that could separate and become gaps.

Looks. Thermal concrete has no esthetic impact; it doesn't change the
outward appearance of a chimney.

DISADVANTAGES

Price. This has historically been the most expensive chimney lining technique, often running $2,000-$4,000 for a routine residential reline,
though recent increases in the price of stainless steel installations
have narrowed the gap. Thermal concrete is also out of the question
for a do-it-yourselfer seeking to save some money: the equipment
for installing it is expensive and generally available only to franchised
dealers.

Hard to clean. Another drawback to thermal concrete is that cleaning it
can, in theory, be a problem. The material is fairly soft, so standard
metal wire chimney brushes shouldn't be used, as they are too abra-
sive. The nylon and polypropylene brushes recommended for these
flues generally will not remove hard, crusted creosote formations.
My experience has been, however, that thermal concrete flues very
rarely contain this sort of hard-to-brush creosote. The Ahrens people
avoid the problem altogether by making a second pouring of a hard
refractory material; the inner surface that results is abrasion-resis-
tant.

Finally, although this is not a disadvantage so much as it is a cautionary note, any chimney lining technique, including thermal concrete, can
be difficult to apply to a flue that is not straight. Offset flues are common in old chimneys, and since it is critical that the flue liner be continuous and surrounded by insulation (in the case of stainless steel) if
dangerous exterior hot spots are to be avoided, an acceptable job must
include the installation of spacers to keep the liner centered. This may
involve breaking into the chimney at the offset, if feasible, or careful
measuring and installation of spacers on the bladder. Either is a pain in
the neck for the contractor but will be done anyway if he or she is
conscientious. If your chimney has an offset, it wouldn't hurt to discuss
with your contractor how it will be dealt with before the work begins.

Installing Flexible stainless steel liner

Stainless Steel

There are two basic types of stainless steel chimney liners: 1.) rigid—stovepipe sections made of stainless steel and secured to each other by pop rivets, installed with a special tool to make a more or less permanent fastening; and 2.) flexible. Absent the need to remove damaged tile liner, either can be installed by a consumer who has no disabling fear of heights, a fair amount of strength, a little ingenuity, a lot of confidence and at least one willing helper with the same attributes. The idea is to put the liner in the center of the chimney, fill the space around the outside with a perlite, cement and water mixture (a special insulating blanket can be wrapped around the pipe in lieu of the mix before installation), and then rain-cap the pipe (to keep out water which will speed corrosion) and flash the space at the top of the chimney between the pipe and the outside edge of the chimney to keep the insulation from becoming waterlogged. This relatively simple procedure becomes more complicated when the flue being lined is offset (see the discussion of thermal concrete) and it becomes necessary to install elbows, if rigid pipe is being used, or spacers, if you are using flexible liner.

The question of whether to use rigid or flexible pipe will be answered by your chimney and your bank account. If your chimney is straight and your wallet is thin, rigid pipe will be your best bet, since it is substantially cheaper than flexible. Offset flues make flexible look more attractive, since it can often be installed without breaking into the chimney, but no matter what the state of your finances, get several estimates; since most chimney sweeps install stainless liners, there is likely to be a fair range of prices available. Be sure to discuss with each bidder what the price includes. How will the flue be insulated? How will he or she deal with offsets (if any)? What does the manufacturer's warranty cover? Does the contractor understand that the liner is subject to thermal expansion, so the anchoring and weather sealing at the top of the chimney must allow for a good deal of natural movement? Is the liner material solid stainless? It should be. It most likely will be Type 304 stainless, which has acceptable resistance to both heat and corrosion. If another type is suggested, find out why and get a second opinion, since different types of stainless steel possess different qualities of heat and corrosion resistance. Also, if the liner is to be rigid pipe, it should be 24-gauge or thicker. Flexible liner must be lighter gauge to be flexible.

ADVANTAGES

Simplicity. It is possible, in some cases, for a homeowner to install a stainless liner, thereby saving hundreds of dollars.

Effectiveness. The finished product is a good liner, warm and securely connected.

Cleaning. Stainless steel liners are comparatively easy to clean.

DISADVANTAGES

Price. Although these liners may be cheaper than thermal concrete, they still aren't cheap, and nothing lasts forever, including stainless steel chimney liners. The manufacturer's warranty, which usually covers the materials contingent upon correct installation and maintenance, will probably be for ten years. In fact, a stainless steel liner may and probably will last much longer—20 years or more—if you take care of it. Conversely, you can ruin it by neglecting it. The inside of a chimney is a very corrosive place, and even stainless steel will deteriorate if it's never cleaned. (This, by the way, is why lining a chimney with standard galvanized stovepipe is a very bad idea, since it will probably fall apart within a year.) Remember also our previous discussion about stainless steel and chimney fires.

Clay Tile

It is sometimes possible to find a mason with the equipment and knowhow to line a standing chimney with flue tiles. This is a tricky proposition, because the only way to seal the joints between tiles is to apply mortar. (It should be refractory cement, actually, which is specially formulated to withstand high temperatures.) The mortar is applied to the top of each tile before it is lowered, and then the tile must be lined up perfectly with the section of tile that preceded it. This can be done with the right equipment, though even then you can't be sure of the integrity of each joint. Done with the wrong equipment, this chimney lining technique is a complete waste of time and money. A case I encountered a few years ago perfectly illustrates my point. The contractor, lacking the inclination to do an honest job, lowered each flue tile with two pieces of wire which he ran under the tile in an X-pattern. Of course, he couldn't remove the wire once the tiles were in place, so he left it, resulting in a flue with two wires crisscrossing its interior at each joint; impossible to clean properly.

Why should you worry about a few gaps in a chimney liner, anyway? The answer is that those gaps can become large holes in a severe chimney fire. Also, creosote can work its way through an unsound joint into the space between the liner and the chimney wall. If it ignites—which it can—it is very difficult to extinguish and can quickly spread to the house.

ADVANTAGES

Price. The main advantage—indeed, the only advantage—of clay tile relining is that it should be substantially cheaper than thermal conrete

or stainless steel and, done properly, is much better than no liner at
all.

DISADVANTAGES

Uncertainty. As noted, it's impossible to know for sure that each joint is
sound.

Tricky installation. It is difficult to maintain the necessary ½-to 1-inch
air space between the tiles and the inside of the chimney shell when
relining.

Also, it may be difficult to find a qualified mason in your area who
offers this service.

When Not to Line

Every chimney flue used with a woodburning appliance (or any other
kind of appliance, for that matter) should have a liner, but this doesn't
necessarily mean that lining your existing chimney is your best option.
Sometimes it is cheaper or more sensible for other reasons to build a
new chimney. Some old chimneys are so badly deteriorated that they
would have to be rebuilt before they could be lined. Or an old chimney
could be unsafe to use for reasons a liner will not correct, such as one I
saw that had a floor joist running through the middle of the flue. Or
perhaps the chimney will need to serve two or more appliances and will
therefore need multiple flues. It isn't always possible to put two flues,
let alone three, in an existing chimney, and even when it is, it may cost
more than rebuilding.

CHAPTER 6

Chimney Cleaning

Burning wood in a stove or fireplace produces good things, such as warmth and esthetic enjoyment. But since wood fires don't result in complete combustion, burning wood also produces an undesirable product: creosote.

What is Creosote?

Creosote is condensed smoke, a highly combustible mixture of tar droplets, vapors and other organic compounds. It always begins as a liquid, but may appear in your chimney in different guises, from flaky and powdery to hard and glazed. Which type of creosote you see in your heating system will depend on the amount of moisture in the smoke, which in turn depends on the system itself, the wood you burn, and your habits of operation. Is your chimney exterior, and do you habitually operate your stove at a low temperature to avoid overheating your house? If so, the odds are good that you will find glazed creosote in your flue, since you regularly create conditions ideal for its formation: low temperature fires that produce lots of smoke, relatively little heat in the flue and, by extention, low draft velocity so the smoke stays in the system long enough to cool substantially.

The opposite extreme is an interior chimney serving an open fireplace in which hot, active fires are burned. The warm flue, unrestricted oxygen supply and strong draft normal to this type of system would probably allow very little creosote to form, and that which did would be flaky or powdery. Between the two extremes are many possible variations: flaky creosote near the bottom of the chimney and no creosote at all near the top; no creosote at the bottom and glaze at the top. The list could go on, and it is this variety which makes a sweep's life so entertaining, if you'll permit me a little joke.

Removing the creosote from a wood heating system is one of the ways in which woodburners pay the piper: it isn't fun, but it's absolutely necessary if the music is to continue. Creosote is bad for your chimney in three principal ways.

CHIMNEY FIRES

Creosote in your chimney can ignite and burn at temperatures well in excess of 2,000 degrees Fahrenheit. This is a chimney fire, folks, and while a number of veteran woodburners start them intentionally to clean their chimneys, the truth is that every time your house survives one intact, you've dodged a bullet.

Chimney fires result in tens of millions of dollars worth of damage annually in the U.S. and Canada. In some instances, the damage is minor, but in many instances lives and entire houses are lost. You may not even be aware of the harm a chimney fire has done, since it is often confined to the interior of the chimney. I'd have to guess that over a third of the chimneys I see after a chimney fire have sustained obvious damage to their flue liners, and I tend to suspect that no chimney emerges from a severe fire completely unscathed.

A case that sticks in my mind is one I encountered a few years ago. The house, perhaps ten years old, was owned by a family that used it for vacations. I was called to do a routine chimney cleaning by the property management firm that took care of the house in the owner's absence. I checked inside first, and determined that there were a fireplace and woodstove sharing a flue. After assuring myself that dust and soot would not migrate from the chimney to the house while I swept, I got on the roof. There I discovered that the chimney had suffered a fire of such severity that large pieces of the clay tile liner had broken away, leaving the chimney completely unsafe to use. Not only had the owners not realized that the chimney was damaged, they hadn't even known that they had had a chimney fire, possibly because it occurred while they were out for the day.

CORROSION

The second way in which creosote harms a chimney is less spectacular than a chimney fire, but a cause for concern, nonetheless: due to its acidic nature, creosote left in a heating system will cause deterioration of masonry or metal.

CHIMNEY OBSTRUCTION

The third harmful effect of creosote is that it can completely block a chimney, due either to the compound's tendency to expand in a chimney fire, or because of buildup during normal (though incorrect) operation of a woodstove. You'll be able to tell if your chimney is blocked, because your house will fill with smoke if you try to use the stove. Removing the blockage can be a great deal more difficult (and costly) than diagnosing it, sometimes requiring that the chimney be opened when less draconian methods fail.

When to Clean

Let's agree, then, that to remove creosote and keep your heating system working as it should, your chimney should be checked and swept, as needed. How often? Every chimney sweep gets asked this question frequently, and the one correct answer—whenever it needs it—rarely satisfies the person who poses it.

It is impossible to answer this question more definitively because not only does each heating system differ from every other, the same heating system will differ from itself from one year to the next: different wood and different weather conditions can lead to rapid creosote buildup in a normally clean burning system. I've had some customers whose chimneys have needed monthly cleaning one year and just two cleanings the next.

So how often *might* your chimney need cleaning? According to the Chimney Safety Institute of America, whenever there are creosote deposits ¼-inch thick—less if the creosote is glazed, since the glaze stores a great deal of energy, and so can produce an especially hot, dangerous chimney fire. The only way to know when it is time to clean your chimney is to check it. For some people, whether because their heating system is difficult to examine or because they wisely refrain from climbing roofs, this will necessitate calling a good chimney sweep, and it should be done before the heating season and again when the sweep advises. Because most chimney sweeps are understandably reluctant to check the same chimney every few weeks, and because most woodburners are understandably reluctant to pay them to do so, I believe that it makes good sense for people to learn to check their own, if it's feasible. This is not necessarily in lieu of professional inspection, but as a first line of defense.

Who can do this? If your heating system is simple—a straight flue venting either an open fireplace or woodstove or furnace—and you have no problem getting to the top of the chimney, then you're in business. Offset flues and difficult systems, such as stoves or inserts installed in fireplaces, are not impossible to check, but because you can't see the whole system, you can't inspect with certain results. If you see ¼-inch of creosote, you'll know that the chimney needs cleaning, but the absence of creosote where you can see doesn't mean that there's none elsewhere.

How to Check a Fireplace Flue

Remember, the flue must be straight, or nearly so, and obviously, I hope, the fire must be out. Dress in old clothes and gloves, and spread a drop cloth in front of the fireplace. Equip yourself with a strong flashlight or drop light, and carefully open the damper as far as you

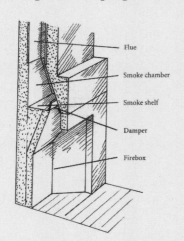

Flue

Smoke chamber

Smoke shelf

Damper

Firebox

can—carefully because forcing some dampers may dislodge them. Before you look up, you should put on eye protection, because you will probably have to stick your head a good way into the fireplace to see up through the flue and will thus be vulnerable to falling soot, creosote and mortar. You will also be exposing your lungs to dust and ash, so wearing a respirator is a good precaution to take.

Shine your light into the wide area just above the fireplace's throat, or damper opening. This is called the smoke chamber. It is usually wider than the actual flue and will not be tile lined. It may be, and should be *parged*—coated with a smooth layer of refractory cement—but more often the smoke chamber will consist of courses of brick tapering up to the flue opening. In an open fireplace, the smoke chamber is usually dirty long before the flue.

Now direct your light into the flue. If the damper opens wide enough and the chimney isn't too tall, you may be able to see all of the way to the top, in which case you can leave the ladder in the garage, if you wish. If you don't have a clear view of the entire flue, then it's up on the rooftop with you. Bring your light, of course, and look down the chimney. On a very bright day, you may have to put your face virtually in the flue to see anything. What you are looking for is creosote, which will be black or dark brown and, in a fireplace chimney, probably dusty or flaky. If you see ¼-inch of it anywhere, it's time for a cleaning. If you are uncertain what you're seeing, call a chimney sweep and, when he or she arrives, ask for some tips on what to look for along with the cleaning. While you're on the roof, by the way, take a moment to check the condition of the chimney's exterior, including its crown (the mortar that fills the space between the liner and the outer walls of the chimney) and the flashing: cracks and separation in these places can lead to moisture damage.

How often you need to practice this dirty ritual depends upon how much you use the fireplace. Their chimneys generally don't need cleaning as often as do woodstove chimneys, so if your use of the fireplace is occasional, you will probably be safe checking it once a year. Heavier use dictates more frequent checking, and so does a fireplace with glass doors, as the doors reduce the amount of combustion air available to the fire and may result in slower, smokier burning. Do get a sweep to make the first inspection and, if necessary, clean the system, so that you're starting with a clean chimney and not leaving major problems undetected.

How to Check a Woodstove Chimney

Forgive me for saying so, but the fire in the stove should be out, or you probably won't be able to see anything but smoke in the flue. You'll need the old clothes and gloves and flashlight. You'll also likely need a small mirror and a drop cloth, and you may need a metal pail and ash shovel, a screwdriver and some kind of respiratory protection.

Start at the base of the chimney—in the cellar, if you have one. What you're looking for is the cleanout, which should be seen as a small steel or cast iron door near the bottom of the chimney. If there's an opening but no door, you need a door, since the chimney will draw air through the opening, ruining the draft and adding an uncontrollable source of oxygen to any chimney fire. If there is no cleanout, you need one (unless the flue doesn't extend to the base of chimney, and the thimble provides reasonable access to the bottom of the flue), since creosote that can only be reached via a cleanout can ignite and, possibly, start a chimney fire. In either case, call a mason or a sweep who does masonry repair. In the meantime, you can still check the flue by removing the smokepipe from the stove and checking through the thimble. A messy job.

If the chimney has more than one flue and more than one cleanout, ascertain which cleanout serves the woodstove flue. Note that if one of the flues serves a fireplace, the woodstove flue is probably offset where it passes the fireplace, making it unlikely that you'll see anything from the cleanout.

OK—the flue is straight and has a cleanout. Put your respiratory protection on, spread your drop cloth and open the cleanout door. You may have to pry gently with the screwdriver. If the cleanout is full of creosote, don't jump to conclusions; the chimney may be clean, since creosote will often fall off the inside of the flue by itself. Shovel the stuff into your pail—slowly, to keep the dust down, and when the cleanout is empty, remove a glove and put your bare hand in (the cleanout, not the pail). You should feel a draft—moving air. If you don't, reach into the cleanout above the door with the shovel or screwdriver and see if more creosote is hanging up out of sight. If so, dislodge and remove it. Now hold the mirror in the cleanout and tip it toward you until you can see all the way up to the sky. A bright day with sun overhead is the best time to do this.

If you can't see the sky, keep adjusting the mirror. If you still can't, either the flue isn't straight or it's blocked at some point, and you should call a chimney sweep.

If you can see the sides of the flue in your mirror, you're looking for the horizontal joints between the tiles all the way up, and clean, smooth tile surfaces. I wince when people say, "I checked it, and it doesn't need cleaning; there's light coming through." You can have light coming through a flue that contains enough creosote to burn a good-sized cruise ship to the waterline. You're not really home free if you see smooth tile, either, because it could be coated with glazed creosote, but if the flue is short enough and the day bright enough for you to ascertain that the tiles are pink or orange all of the way to the top and not brown or black, you're headed in the right direction and can probably forgo the trip up the roof, unless the view from up there on such a fine, bright day is enough to lure you up to check the condition of the masonry, as discussed earlier.

If you aren't certain whether the chimney is clean or dirty, head for the rooftop with your flashlight. Sometimes putting a drop light in the cleanout gives a very good view of the chimney's interior from the top and, always, if you have enough extension cord, lowering a drop light slowly through the flue will give you a good idea of what's there. As with the fireplace flue, if you have any doubts concerning what you're seeing, call a chimney sweep. And do get a sweep to check and clean your chimney the first time. You can check it yourself when the sweep is done and thus establish a benchmark for your future inspections.

The last thing to check is your stovepipe. Some people try to get away with tapping the pipe with a fingernail—a clear, hollow sound being indicative of a clean pipe—but certainty can only be attained by taking the pipe apart and looking inside with a light. Check the thimble, too; creosote often builds up rapidly at this point in the system, particularly if the thimble is a foot or more long. This is also a good time to assess the condition of your stovepipe. Is it free of rust and fairly rigid? Is each joint secured by at least three sheet-metal screws so that the pipe can't shake apart in a pipe fire (same as a chimney fire, but confined to the pipe, at least temporarily)? Again, it's best to go through the procedure the first time with a qualified chimney sweep.

Sweeping Your Own

I'm sure that some chimney sweeps will disagree, but I believe plenty of people are well advised to sweep their own chimneys, particularly if they can secure the cooperation of a competent sweep who is willing to check their work from time to time. Who are these people? For the most part, they'll be among the people who can check their own chimneys, excluding those with fireplaces: cleaning a fireplace chimney is obviously not impossible—I can do it, after all—but it does require equipment beyond brushes and rods, and it can be a filthy undertaking, since part of the work must be done in the fireplace with the damper open. The sweep can certainly get dirty, exposed as he is to a shower of creosote as he cleans the smoke chamber. The room in which the fireplace is located and, in extreme cases, the entire house, are also in danger of becoming soot covered unless a special vacuum cleaner with great filtration and air processing capacity is employed.

Why would anybody choose to clean his or her own chimney? With prices for a professional cleaning averaging, at this writing, $65-$100 per flue, many people of modest means who need several sweepings a year might be tempted to forgo some or all of them; better that they should hire the sweep for one and learn to do the rest themselves. Even woodburners who have no intention of doing this particular dirty work will, I believe, benefit from having some basic knowledge of the process of chimney sweeping; not only will you be better able to evaluate the pro's work, you'll probably find that your understanding of your system and its components is enhanced. This is not a part of life that benefits from mystery.

Alternatives

Before I explain how to sweep your chimney, I'll briefly discuss three other apparently inexpensive sources of chimney cleaning.

CLEAN-IT-YOURSELF RIGS

These have the advantage that they don't require you to climb the roof. The ones I've seen consist of a frame and pulley at the top of the chimney, a brush in the cleanout, and a loop of cable which is attached to the brush and runs through the pulley at the top. By pulling on the cable, you can work the brush up and down through the flue in the comfort of your basement. These can work, but they have four major problems:

Obstruction. The cable in the flue interferes with the passage of smoke and contributes to creosote buildup.

Short life. The acidic nature of creosote, exacerbated by the condensation in the flue, breaks down the apparatus fairly rapidly.

Incomplete cleaning. If your chimney is inspected and swept exclusively from the basement, it's conceivable that glazed creosote in the system could go undetected, and since your brush will not remove it, you will be vulnerable to an intense chimney fire.

Pollution. Cleaning a chimney through an open cleanout may expose you to large   amounts of carcinogenic dust and gases, and your house to a mess you'll not soon forget. It can be done safely and cleanly, but not easily and not without a good respirator and a good vacuum cleaner.

LOCAL FIRE DEPARTMENTS

The second chimney cleaning source I have in mind isn't as common as it once was, but may still be available in some areas. Volunteer fire departments will sometimes sweep chimneys in exchange for a dona-tion. This seems innocuous enough, and sometimes is, but it can also lead to real problems. Firefighters may know a great deal about extin-guishing fires, but there is no reason to suppose that they know any-thing about chimneys or chimney cleaning.

Two cases illustrate my point. I was called to clean a chimney that was malfunctioning; whenever the homeowner lit the stove, smoke would pour into the house. When I arrived, I found a stove installed in a fireplace, with the pipe passing through the damper and a metal plate closing the remaining space in the damper opening. The customer explained that she was puzzled by her suddenly smoky stove, since the local fire department had cleaned the chimney just a week before and had, in fact, cleaned it the preceding four years. On checking, I found the flue was not really clean, but neither was it dirty enough to restrict the draft. When I pulled the stove out of the fireplace, however, I discovered the source of the problem: the firefighters had evidently pushed the creosote down the flue year after year but had never ques-tioned where it was going, which was, of course, into the smoke cham-ber and stovepipe, both of which were packed full. By the time I was done, I had filled three 30 gallon garbage bags with creosote.

The second case involved a department that offered a safety inspec-tion of any local resident's wood heating system. A very well-meant offer, but potentially a very expensive one, since the firefighters did not know the acceptable clearances from combustibles for stoves and stove-pipes. One installation I saw, which had been approved as safe by the firemen, featured a woodstove installed in a fireplace a scant foot below a wooden mantel. The distance should have been three feet. I have heard of other departments whose members cleaned chimneys with chains or feed bags full of bricks, either of which can break flue tiles. All in all, I think that you'd be well advised to let the firefighters stick to firefighting, give them the donation and thanks that they deserve for doing that dangerous and vital job, and have somebody who is qualified sweep your chimney. The problem that the firemen face is that you need to know a lot about chimneys and wood heating and have considerable experience with both to recognize problems in heating systems with which you are unfamiliar. You may be able to learn enough about your own system to do it justice, but going beyond that involves the sort of commitment to learning the trade that a good chimney sweep makes.

CHEMICAL CLEANERS

The third cheap source of chimney cleaning are the various chemical cleaners on the market, such as Rutland Liquid Chimney Cleaner, Safety Flue or Anti-Creo-Soot. These are usually either liquid or powder that you add to the fire. The problem is that they don't clean the chimney. Some may help convert glazed creosote to a brushable form of the menacing stuff and, in the process, cause some of it to fall off of the sides of the flue, but chemical cleaners are not a substitute for manual cleaning under any circumstances.

Sizing Up the Job

Now you're ready to learn to sweep your own chimney. If you thought that inspecting it was a nasty business, you're in for a real treat now.

I'm assuming that your system consists of a woodstove connected to the chimney through a thimble, rather than through a fireplace. I'll add that, even if your stove does connect via a thimble, if it shares a flue with a fireplace (the thimble being either over the mantel or on the side of the chimney opposite the fireplace), you should leave the job to a pro, because the smoke shelf will function as the cleanout in this case, and the removal of the creosote can be tricky and will probably require a special vacuum cleaner.

I'll also acknowledge that a simple woodstove system can often be cleaned by a homeowner even if it has an offset flue; you may not be able to inspect it, but with the right equipment, you can sweep it.

EQUIPMENT

A chimney brush sized to your flue. You'll need to measure the inside dimensions of your flue. Lined chimneys are standard nominal sizes: 6-, 7- and 8-inch round, 8- by 8-inch and 12-by 12-inch square, and 8-by 12-

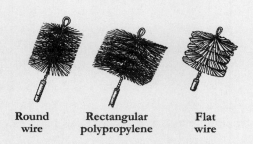

Round wire   Rectangular polypropylene   Flat wire

inch and 12- by 16-inch rectangular tiles are very common. Be aware that these measurements are exterior in the cases of square and rectangular flues, however; the interior dimensions will vary somewhat, but will typically be diminished by an inch in each direction—8- by 8-inch would be cleaned by a 7- by 7-inch brush, for example.

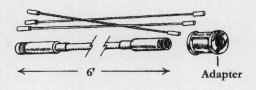

6'   Adapter

Chimney brushes come in round wire and flat wire as well as polypropylene and nylon. For most people, the round wire brush is a good bet; it's less expensive than the flat wire models and will do an adequate job of removing powdery and and dry, flaky deposits. If your system tends to generate hard, scaly creosote, however, as opposed to hard, glazed creosote, you'll need to pony up the money for a flat wire brush; its stiff, chisel-like tines are effective in chimneys a round wire brush can't touch. If a flat wire brush is unable to remove the gunk in your chimney, it's past time that you called a sweep. The nylon and polypro brushes are the least expensive, but will only work on very dry, dusty deposits. Fireplaces and some insulated metal chimneys are often well cleaned by these brushes.

Rods or rope: Most brushes have a ring at one end and a threaded stem at the other. Chimney sweeps almost uniformly use flexible rods which attach to the stem of the brush and then to one another in series until the brush has traveled the length of the flue. Adaptors are available that allow you to attach a ring to the stem and use the brush with a rope instead of the expensive rods, either by hooking a rope to one ring and a weight to the other and lowering the brush down the flue, or by tying ropes to both rings and having a helper at the cleanout pull while the brush is lowered from above. The rope method can work, but be aware that, if you use a weight instead of a helper, you run the risk of cracking flue tiles, particularly in an offset flue. The helper method has a significant drawback, too: it requires that the cleanout door be open during cleaning. We've discussed the negative impacts of that already. I tend to think that you are better advised to invest in a set of rods: they'll last you forever, unless you become a professional chimney sweep, and will amortize themselves in a couple of cleanings.

The most commonly used chimney rods are made of fiberglass. They come in three diameters—.480-inch, .440-inch, and .340-inch—and in lengths of 3-, 4-, 5- and 6 feet. longer rods are cheaper by the foot, and shorter rods are usually handier for cleaning from the bottom of the chimney. Most sweeps rely on the heavy-duty .480 rods because they have enough rigidity to push a brush through a substantial creosote buildup. I think, however, that most homeowners will be well-served by the .440 rods: they cost less, and if you find that they are inadequate for your chimney, you are either waiting too long between cleanings or should be hiring a pro, or both. The .340 rods often work, but they're quite noodly and have difficulty moving a brush through more than a very light buildup. Their usual application is cleaning from the bottom of the chimney, which we'll discuss presently.

A shop-type vacuum cleaner is optional. If, for example, your stove and cleanout are in a less-than-pristine part of the basement, you may find that a broom and dustpan are an adequate alternative. If you have to deal with creosote in an inhabited part of your house, however, you'll want a vacuum and are ill-advised to use your household model, since the very fine and corrosive dust that often comes out of a chimney may harm the motor.

A pail—preferably with a secure lid—and an ash shovel for removing creosote from the clean- out. You don't know what trouble is until you've had a paper or plastic bag full of creosote rip halfway across the new white shag carpet in the living room.

A short-handled wire brush for cleaning the stovepipe. If you have more than 2 or 3 feet of pipe, you might consider buying a round chimney brush to fit it: attached to one length of rod, it will do a fine job of cleaning the pipe and will make it unnecessary for you to reach way in with the hand brush and fill your gloves with soot.

Whatever *screwdrivers* are necessary to disassemble and reassemble your pipe. The most difficult and frustrating part of most chimney jobs is getting the screw holes in the pipe to line up when you're putting it back together. To minimize damage to my sanity, I disassemble the pipe only enough to remove it and clean it, and make a light mark overlapping any joints I do separate so that I can more easily match them up. I've also found (because my wife told me) that an awl makes short, easy work of lining up screw holes that don't quite want to go together.

A note on cleaning pipes: because taking them apart can be such a hassle, some people bang on them, instead, using their hands or a stick. It is true that this forceful approach will often yield the satisfying sound of creosote falling through the pipe, but there are several problems with the technique. Where does the creosote go if you don't remove the pipe from the stove? If there are no elbows in the pipe and no baffles in the stove, it will probably fall into the firebox, which is okay, but remember: you haven't removed the creosote, you've just consolidated it at its angle of repose. Also, hitting the pipe usually doesn't loosen all or nearly all of the creosote; even if it falls into the firebox, you've probably left enough in the pipe for a good fire. Consider also that if you don't remove the pipe, you don't inspect the thimble, which is probably dirty. Finally, banging on the pipe may make it fall apart.

Personal protective equipment: Creosote is a carcinogen and skin contact and breathing it are to be avoided as much as possible. Obviously, cleaning your own chimney a few times a year will give you very limited exposure, and you will probably be reluctant to spend hundreds of dollars for a powered, positive pressure respirator, but you should consider a nonpowered respirator with cartridges rated for both dust and vapor. It isn't just creosote dust that is dangerous; the vapors are carcinogenic as well, and this is why paper dust masks are inadequate.

Clothes that fit snugly at the sleeves and neck (turtlenecks—black turtlenecks—are perfect), and nonabsorbent work gloves and a hat should provide a reasonable amount of skin protection, particularly if you make it a point to shower and change clothes as soon as you've finished sweeping.

Drop cloths are also optional, but spread around your indoor work area, they'll save you a lot of cleanup time.

A good *flashlight* or a drop light with plenty of extention cord. I use both, but if I had to pick one, it would be the flashlight for its handiness.

Ladders. It is not my intention to give a complete course on ladder use and safety—I'd have trouble passing such a course, anyway. I'm going to assume that, as I mentioned earlier, getting on your roof is no problem for you. If you've never done it before, I don't want to goad you into trying, particularly if your roof is slate with a 12/12 pitch atop a three-story Victorian. In lieu of the whole curriculum, I'll give you my favorite tips about working on roofs, minus a few that it would be irresponsible of me to pass on.

Working on the Roof

Unless your roof is flat or very modestly pitched, get a ridge hook and attach it to a single section of ladder long enough to reach from the eaves to the peak. Walking a ladder beats walking a roof any day.

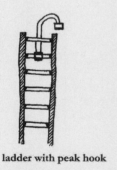

ladder with peak hook   peak ladder in place

Some roofs will not accommodate a peak, or ridge ladder either because the distance from eave to peak is too great or because they don't have ridges (*e.g.*, a hipped roof). This may be enough to scrub the do-it-yourself project unless your chimney can be cleaned from below, but if you think that you can walk the roof, I will, though with trepidation, rate the various common roofing materials for their walkability:

Asphalt and fiberglass shingles usually offer good traction, even when wet. They can be icy and not look it, though, so proceed with caution in cold weather.

Metal roofs are unpredictable: old galvanized roofs are often walkable, and so are enameled raised-seam roofs if they aren't too new and you can help yourself up by grabbing the seams. Avoid them in damp weather.

Slate is usually fairly slippery and is fragile, besides. I put a peak ladder on slate roofs whenever it's possible, and walk carefully, near a gable end, holding the edge with one hand, when it isn't. Don't try to walk them when they are wet.

Wooden shake roofs are much like slate, but are even more slippery when wet. They are also difficult to slide peak ladders on, because the thick edges of the shakes catch the top of the ladder. If you are walking a dry shake roof, watch out for moss and lichen growing on it; they hold moisture and will be slippery when the rest of the world is drought-stricken. Why do people have shake roofs?

Don't lean a ladder on a freestanding chimney, even a new one. Mortared joints are intended to hold the chimney together, not to provide great lateral strength. Metal chimneys won't even provide the illusion that they can support you. When a chimney extends too far above the roof to be reached, I stand an aluminum step ladder against the chimney—not leaning, but vertically—and tie it to the chimney. I don't even do this when confronted with a tall metal chimney, however; I either take the chimney apart or clean it from below.

Have somebody foot the ladder when you climb. Many sweeps work alone and devise ways of using ladders without assistance—driving stakes and tying the bottom rung to them, for instance—but what working alone often means is taking chances.

When walking a roof, keep your weight centered over both feet by taking short steps, and keep a low center of gravity by staying crouched—the steeper the roof, the deeper the crouch. Holding the edge of the roof or even just putting your free hand flat on its surface will help keep you from slipping.

You aren't climbing the roof for the fun of it, so you'll have to carry rods and a brush. It's always nice and often essential to have a free hand, so devising a way to carry your tools one-handed is necessary.

Attaching the brush to one of the rods and then tying all of the rods in a tight bundle would work.

The best shoes for roof walking are said to have gum-rubber soles, but I find that old running shoes work well, too; it's a good way to recycle them after their pronation control is shot. (I'll hold off on identifying my favorite brand here in anticipation of an endorsement contract.)

Where do you keep the extra chimney rods when you're cleaning the chimney? I usually lay them across the top, between me and the flue. They aren't very likely to roll off, because they'd have to either go through me or the rod currently sticking out of the flue. It is a good idea, nonetheless, to keep people, cars, furniture and any other breakable objects a safe distance from the eaves in case something falls.

People assume that chimney sweeps use safety ropes, and some do. The problem is that there isn't much on a roof to which you can tie yourself; the chimney is probably it, and some chimneys would not support you if you fell, but would add to your considerable troubles by falling on you shortly after you landed. If you are going to use a rope, tie it to the base of the chimney rather than the top or middle; it's less likely to fail if your weight is suddenly suspended from the chimney.

It's easier to climb up a roof than down, for some reason, perhaps because on the way up your direction of travel tends to nullify the subversive influence of gravity. If you suspect that the ascent will tax your nerve, reconsider the whole project; the descent will likely be beyond you, and you could get awfully cold, bored and hungry sitting on your roof waiting to be rescued by the fire department.

Cleaning from the Roof

I'm going to describe the steps in a routine top-down cleaning. This assumes that the chimney is not dirty enough to pose technical problems: no glaze or heavy, compacted buildup that would require special equipment to remove.

PRELIMINARIES

Secure your working area. Before setting up your ladders, you should identify all openings in the flue you're going to sweep. There should be

only two—the thimble and the cleanout—but, particularly in older houses, there may be more, and they may be hidden. I could tell you a nightmarish story about an open thimble behind a painting in a restaurant, and I've heard of open thimbles behind wallpaper that revealed their presence catastrophically when the chimney was swept. If you aren't positive that you know all of the points where your chimney is breached, examine it as thoroughly as possible. Incidentally, any openings that are unused should be sealed with masonry to the thickness of the chimney and tile walls: in the event of a chimney fire, such openings can allow the fire to spread to the house and also provide an uncontrolled source of oxygen for the fire. Those tin pie plates commonly used to plug old thimbles have about as much chance of standing up to a 2,000-degree chimney fire as would a wad of paper towels.

Now that you've identified the openings in the chimney, make certain that they are sealed. A thimble with a tightly fitting stovepipe in it is often secure as is, but to be on the safe side, tape the joint where the pipe meets the thimble with duct tape. If your chimney is metal, tape the joint where the stovepipe meets the chimney.

Cleanout doors often aren't very tight and don't latch securely; tape between the door and its frame. Unused thimbles can be tricky to mask, particularly if they don't protrude from the face of the chimney.

If tape will adhere to the chimney surface, you can put a plastic bag or crumpled-up newspaper in the thimble and tape a sheet of plastic over the opening. If you can't tape over the opening, pack the thimble well with paper or plastic, being careful not to push it into the flue, and brush slowly to avoid creating a sudden vacuum which could dislodge your barrier.

One last precaution should be taken if your chimney has more than one flue: not all the creosote falls to the bottom when a chimney is swept; much of the very fine stuff comes out the top. There is the possibility, if you have flues adjacent to the one you are sweeping, that some of the dust will go down and get in your living space through openings in that flue. Fireplace flues, probably because of their large diameters, are particularly susceptible to this problem, so before you start brushing, close the damper and, for extra security, consider covering the top of the flue with a board or plastic bag while you are working on its neighbor.

Now to the roof. I like to make one trip to the chimney without my tools, mostly to establish my route and identify obstacles (slippery flashing at the eaves, inconveniently located TV antennas and lightning rods, wasp nests, etc.), but also to check the condition of the chimney. When you've checked the roof and chimney, get your tools ready to carry. Don't forget a flashlight to help you check your work, and don't forget your respirator; despite being outdoors, the top of a chimney while cleaning is in progress is an atmosphere choked with toxic gases and vapors. Up you go.

Cleaning the flue. To clean, insert the brush, with a rod attached, in the flue and push it slowly to the bottom of the chimney, adding rods as needed. You may hear creosote falling, and dust and soot will likely begin to come out of the top of the chimney. If you encounter serious resistance, don't try to force the brush past or through it, since, if it is creosote, you may compact it and form a blockage. Instead, pull the brush back up the flue a few feet and gently try again to brush past the sticking point. (Incidentally, it isn't always easy to reverse a brush in a flue; I've had to rig a jack to accomplish this a few times by laying a stout stick across the top of the chimney, tying the rod tightly to it, and then using the stick as a lever.) If this fails, pull the brush out, give the dust a few minutes to settle, and use your flashlight to determine, if possible, the nature of the obstruction. If it is creosote, you can try, very carefully, to dislodge it with the tip of a bare chimney rod. Failing at that, or if you fail to identify the obstruction, you'll have to call a chimney sweep.

If you encounter no resistance, push the brush to the bottom of the flue and then pull it back out and repeat the process until the chimney is clean, which you can probably determine with your flashlight. Note that clean chimney tiles may still be black, but should be free of deposits in excess of ¼-inch thick. Also, if the inside of the flue is black and shiny, you have glaze and need to get a sweep.

Cleaning the stovepipe. Back in the house, spread your drop cloths around the stove, set up your vacuum cleaner, bucket and hand tools, disassemble the stovepipe and remove it from the stove. Now comes the judgment call: it's usually preferable to clean the pipe outside, but perhaps not if doing so entails carrying it across that white shag carpet. You can try wrapping the pipe in a drop cloth for the trip, but be careful. If you elect to clean it inside, turn your vacuum on and position the hose at the top of the bucket where it will catch airborne dust. Slowly brush the pipe into the bucket—slowly because speed creates turbulence and flying soot—until the deposits are gone. Now clean the thimble with the hand brush, vacuuming to control dust, and clean and vacuum the stove collar (*i.e.*, where the pipe enters the stove), baffles and smoke passages, if any. Be careful about vacuuming any part of the stove if it has been running recently; embers can last an astonishingly long time—three days or more, under the right circumstances. Reassemble and reinstall the stovepipe and, after carefully folding your drop cloths and shaking them clean outside, head for the cleanout. By the way, it is often a bad idea to open windows or doors while the inside work is in progress; the fresh air may seem helpful, but the drafts it may create could disperse fine creosote dust through the house.

Cleaning up and checking your work. To the cleanout bring your drop cloth, vacuum cleaner, pail and shovel, respirator and inspection mirror. Set the vacuum up, start it, and hold the hose slightly above the cleanout door, so the vacuum will help control the dust you generate as you open the door and shovel the creosote into the pail. It may be tempting to leave the creosote in the cleanout if you're in a hurry, but don't do it; a spark could ignite it, and damage to the cleanout door and chimney could result.

Now inspect your work with a mirror. You should see a clear flue all the way to the top of the chimney, unless the flue has an offset, in which case you should feel for a draft. If you have neither draft nor view, you've possibly left a blockage of creosote in the flue, and you'll need to return to the roof unless you can reach and remove it from the cleanout. This is a frustrating experience and one you shouldn't have unless you clean your chimney too infrequently. If I have any doubt that I've brushed clear to the cleanout, I push the brush down as far as I can, then I get off the roof before finishing the cleaning and check the cleanout to make certain that the brush is there.

Finishing. Your last step is to double check: Is the cleanout door tightly closed? Is the stovepipe properly reinstalled with three screws at each joint? Are any repairs needed? OK. You're done. It wasn't so bad after all, was it?

Cleaning from the Inside

Many chimneys can be cleaned from the inside or, more accurately, from the bottom. Fiberglass and polypropylene rods are flexible enough to push a brush through most fireplace damper openings and many cleanout doors and stovepipe thimbles, and since this method will keep you off of the roof, why not use it if you can?

The answer is that without a special vacuum cleaner, the dust generated by from-the-bottom chimney sweeping is very difficult to control. This won't matter too much if your cleanout door is outside, and it may not matter if you can effectively mask the work area, but unless your roof is nearly impossible for you to climb, cleaning from the top will probably be easier. Also, it is important for you to periodically inspect the condition of the chimney from the roof line up, and it is often necessary to clean chimney caps, particularly those on metal chimneys, since they seem to have a penchant for collecting creosote. Having said this, I'll acknowledge that your situation may demand that routine cleanings be done from the bottom.

How can you tell if it's possible in your case? By looking, first, then trying. A cleanout which is not elevated above floor level will usually be impossible to start a brush through, even if you use the light .340-inch rods, because getting the rods into the flue would demand more flexibility of the rods than they could have and still push the brush effectively. The same is true of a flue that is recessed more than a few inches from the cleanout or thimble opening.

If the only contra-indication to cleaning from the bottom is the elevation of the cleanout, however, you may be able to clean through the thimble, if the brush will fit. In any case, you won't know that it is possible to clean from the bottom until you try to start the brush in the flue. This is seldom easy to do; you'll probably have to reach into the cleanout or thimble and push the bottom of the brush up as you apply pressure with the rod.

If you succeed in starting the brush up the flue, here's the procedure for finishing the job:

1. *Secure all chimney openings* save the one you're going to use. You'll be much happier if you can mask that one, too, since a lot of dust will probably billow out of it otherwise. One way of doing this is to start the brush and rods, then tape a piece of plastic or heavy paper with a slit cut for the rods over the opening.

2. *Move cars, people, lawn furniture, etc.* away from the eaves in the proximity of the chimney. Large amounts of creosote may be pushed out, and sometimes loose bricks from the chimney top can be dislodged: both are subject to the laws of gravity.

3. *Set up your vacuum* and drop cloth at the cleanout, if that's where you are working, and position the vacuum hose near the opening.

4. *Start the vacuum* and push the brush slowly up the flue, adding rods as needed. Retract and repeat. Resistance should be dealt with as discussed above, but there is less concern about compacting creosote when cleaning from the bottom, so your main worry is that the resistance is a broken tile or an offset in the flue, in which case you could do damage by applying excessive force.

5. *When the chimney is clean*, follow the procedure previously described for dealing with the stovepipe and cleanout.

Hiring a Pro

If you can't clean your chimney, don't want to clean your chimney or are having a problem with your chimney, it's time to hire a chimney sweep—past time, perhaps. And, just as you need to be selective in choosing a doctor or an auto mechanic, you should have more criteria for picking a sweep than price. To expand upon this comparison a bit, you will derive the most benefit from a good doctor who knows your health history, or a good mechanic who knows your car, or a good sweep who knows you and your heating system. A sweep who sees your chimney year in and year out will be able to spot changes and diagnose performance problems more quickly and accurately than will somebody you hire on a one-shot basis.

Here are some guidelines for picking a chimney sweep—an important decision, after all, since it might well affect the safety of you, your family and your property.

Shop around. Ask other woodburners in your area whom they use. If you live in a small town or rural area, your choices may be limited, but most people in the woodburning belt will be able to find several—and perhaps many—chimney sweeps from whom to pick. Getting some consumer reactions is a good first step.

Look in the Yellow Pages. This will give you a pretty good idea of how many sweeps work in your area, and the ads themselves may reveal or imply some important information: is the sweep certified by the Chimney Safety Institute of America? What services does he or she offer? Is the ad informative, or does it emphasize sizzle rather than steak— top hat and tails rather than service.

Credentials. Let's return to the question of certification. In order to be certified by the Chimney Safety Institute of America (CSIA), the educational arm of the National Chimney Sweep Guild and the only national organization in the U.S. that certifies sweeps, a candidate must pass a rigorous written exam based upon a manual prepared by the CSIA and upon safety codes developed by the National Fire Protection Association. Certification does not guarantee that a sweep will have the mechanical proficiency or inclination to clean your chimney, however, nor is it a guarantee of honesty. And the lack of certification is not proof positive that a sweep lacks essential technical knowledge. What certification does indicate is that the sweep has mastered a comprehensive body of technical information and, moreover, that he or she cares enough about the trade to spend a good deal of time and money to study for and take the exam. While I acknowledge that there are excellent sweeps who are not certified, I would advise you to hire one who is; the trade is not regulated, so the CSIA's certification program provides the only standard besides reputation and personal contact that a consumer can apply in making a choice.

Ask questions. When you've got a short list of prospective sweeps, it's time to make some phone calls. You'll want an estimate of cost, certainly, and you might also ask what the sweep's cleaning method is. If it's scrubbing the flue with a Christmas tree or a live goose, you've heard enough and can hang up with no fear that you are missing out on a bargain. Wire brushes and vacuum cleaners are the standard tools of the trade, and any competent sweep will have them.

You also need to ask what you'll get for the quoted price. Does it include cleaning the stovepipe? If you have a fireplace, does the the sweep clean the smoke chamber and behind the damper? (Don't hire one who doesn't.) Some sweeps charge extra for answering questions—consultation—so you should establish whether or not this is the case before you hire.

Insurance. Be sure to ask whether the sweep has liability insurance, and don't hire anyone who doesn't. If you have any doubts, ask the sweep to furnish you with a certificate of insurance.

I've spent a lot of time on the subject of chimney cleaning because it is critically important to anyone who burns wood. I can't stress enough that wood heating is an active pursuit, not a passive one, and that even if you don't clean your own chimney, learning as much as you can about the process will serve you well.

CHAPTER 7

Troubleshooting Chimney Problems

For all its pleasurable aspects, heating with wood can lose a bit of its luster when problems develop; when, for instance, your stove won't burn hot and fills your house with smoke whenever you open the loading door. Any kind of heating system can malfunction, of course, but wood-fired systems, because of their variables, can pose problems that are not obvious as to cause or solution. Sometimes—often—dealing with such problems is best done with professional help, but there are many cases where a homeowner can diagnose and remove the cause of the troublesome stove.

The first thing to recognize is that, in a vast majority of cases, a performance problem in a woodburning stove or fireplace is a problem with the chimney. An elementary understanding of how chimneys work should help you figure out why they sometimes don't.

How a Chimney Works

A chimney that is functioning as it should is a column of warm air (or gas) which, as it rises, causes a condition of lower pressure—*i.e.,* suction—at the bottom of the flue relative to the pressure in the room where the stove is located. This suction draws air through the appliance where it is heated, causing it to rise and so perpetuate the condition. The suction is known as *draft*. Most chimneys will have draft even when their appliances aren't operating, because the air in the closed column will still tend to rise, and since the pressure at the bottom of the chimney will usually be lower than the pressure in the room outside the chimney. Nevertheless, the hotter the gas in the flue, the faster it rises, and the greater the resulting draft.

Besides containing a column of rising gases, a successfully operating chimney will be able to handle a sufficient volume of gases. In other words, the chimney that vents your fireplace must be large enough to move all of the smoke produced by the fire up and out of the house. Volume is not independent of draft: to a point, the greater the draft, the greater the volume of smoke the chimney can process.

Diagnosing Chimney Problems

Because of the variables in wood heating systems, it's important that you collect data when a problem develops so that whoever deals with it has a basis for narrowing the field of possible causes. Four questions you need to ask and answer include:

1. Did the problem just manifest itself, or is it chronic?
2. Is it constant (the stove always smokes)?
3. Does weather seem to affect it? What weather?
4. Does the stage in the burn cycle seem related to the problem?

Now we'll apply what we know to some sample chimney problems, both short-term and chronic.

SHORT-TERM PROBLEMS: SMOKING

By a short-term problem, I mean one that recently developed in a usually trouble-free system. The most common of these is more or less sudden smoking when the stove is in operation, particularly when the loading door is opened. Most often the cause is creosote buildup sufficient to reduce the chimney's volume capacity to the point where it can't handle the smoke produced by the stove. Frequently, the chimney will actually be blocked in one or more places. An attentive stove operator, particularly one who habitually checks his or her own chimney, would usually notice this problem in an early stage, but creosote buildup can be very rapid under the right circumstances. The immediate solution, of course, is to sweep the chimney. The long-term solution involves finding out why the system is subject to such heavy creosote buildup and trying to remedy the situation. These causes might include:

Sizing. The stove is too large for the space it is heating, and consequently is usually operated without enough oxygen for an efficient burn.

Weather. Relatively warm winter weather can help make chimneys become dirty rapidly. People slow down their fires under these conditions, leading to incomplete combustion and a smaller than usual temperature difference between the gases in the flue and the outside air, which results in decreased draft. A solution is to build smaller fires and not overly restrict the oxygen supply. You might also open a window or two to cool off the house, or you might simply not use the stove during warm weather.

Bad fuel. Wet or unseasoned wood tends to burn incompletely. The solution is to either find dry wood or take pains to burn the wet wood hotter.

In a relatively small number of cases, a stove may start smoking for reasons other than a creosote blockage. These might include:

Weather. Again, warm weather—particularly if it's cloudy, warm and damp—leads to sluggish draft. I've encountered stoves that smoked but had clean chimneys and pipes and worked fine in colder weather. There are likely to be other factors involved in a case such as this, but operating a woodstove in 50-degree Fahrenheit weather can lead to trouble.

A mechanical problem. Perhaps the catalytic converter is plugged (in which case the stove should work fine in its noncatalytic mode) or a broken chimney tile is obstructing the flue, or a chimney cap is plugged. I once responded to a call from a customer with a smoking stove. Puzzled at finding the chimney and pipe clear, and seeing no obvious reason for the problem, I finally removed the top of the stove and found that a sheet of paper—kindling from an earlier fire—had been drawn into the baffle where it blocked the smoke's path to the chimney.

RELUCTANT BURNING

A second short-term problem is a stove that has started to burn sluggishly. The chances are good that this is the same problem as the smoking stove but at an earlier or less severe stage. It could, however, signal a dead catalytic converter, a worn-out thermostat on a stove with a thermostatically controlled draft, or even an excessive buildup of ashes inside the stove which restricts the air intake and circulation.

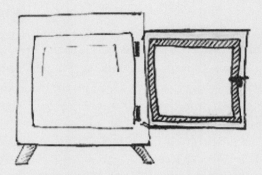

channels in stove door for rope gasketing

TOO FAST BURNING

A third problem is the opposite of the second: a stove that suddenly burns too hot and is difficult to control. A likely cause is overly dry wood which needs very little of the fire's energy to drive off moisture. The solution is to mix the super-dry wood with less dry wood, rather than reducing the fire's air supply, since dry wood in an oxygen-starved fire produces a lot of smoke and creosote.

A second possible cause for a runaway stove is worn-out stove door gaskets. This condition allows the stove to draw air on demand, even when the air inlets aren't wide open, and will result in very hot fires. A loading door that latches very easily and/or has any play in it when closed often has worn-out gaskets. The solution is to replace the gaskets.

CHRONIC PROBLEMS: WIND MAKING SMOKE

Long-term problems are often more difficult to diagnose and more expensive to solve than short-term problems because they frequently are indicative of a heating system with mismatched components or ill-conceived design.

A common problem is a stove or fireplace that smokes during windy weather. This may be the result of turbulence at the top of the chimney, or because the chimney is on the side of the house hit by the prevailing wind, resulting in higher pressure at the top of the chimney than in the house, reversing or negating the draft. In either case, increasing the height of the chimney should solve the problem—in the first case because it would bring the top of the chimney above the turbulence, which is commonly caused by nearby, taller objects (the roof, trees, other chimneys), and in the second case because the raised top will be above the wind-created area of high pressure. A chimney should be at least 2 feet taller than any object within 10 feet of it, and should project from the roof at least 3 feet. These are minimum specifications, though, and may not be sufficient in some cases.

If the smoking is caused by turbulence, a chimney cap may solve the problem inexpensively by deflecting wind blowing down on the top of the chimney.

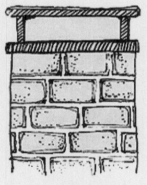

**chimney with
masonry cap**

**chimney with
stainless steel cap**

Smoky Fireplaces

If you have a fireplace that always smokes, the problem is unlikely to be wind. If you are able to eliminate high pressure at the top of the chimney as the source of your woes, consider the following common problems:

Cold Chimney. The flue is too cold and hasn't draft enough to be "self-starting." This situation will usually be associated with an exterior chimney and is sometimes easy to diagnose. If the smoking stops once the fire is well established, and you can overcome the smoking by holding a flaming torch above the damper opening prior to starting the fire, a cold chimney is your culprit, though "stack effect," which we'll discuss shortly, may present similar symptoms and may, in some cases, be part of the reason that the flue is cold. In any event, if the flaming torch works, the solution to this problem is to always use it. Don't shoot the architect.

The flue is too small for the fireplace opening. Again, this is fairly easy to diagnose. Get a board or a piece of cardboard as wide as the fireplace opening, and kindle a fire. When it begins to smoke, hold your board just below the lintel (the top of the fireplace opening) and gradually lower it: if the smoking stops, you may have found the cause. To confirm the finding, measure the flue (see note above, page 56 about measuring flue area) and determine its area, likewise the fireplace opening. The flue's area should be at least one-tenth that of the fireplace opening. If it isn't, your solutions include lowering the lintel permanently or installing glass doors. Don't shoot the mason.

Inadequate draft. The fireplace doesn't have enough air available to maintain the necessary flow. Tight, modern houses in particular often don't allow enough air infiltration to feed a hungry fireplace. To diagnose this cause, kindle a fire. When it begins to smoke, open a window near the fireplace. If the fireplace stops smoking, insufficient air is the likely problem. The solutions are to always open the window, install outside air returns or install glass doors, which, by reducing the effective area of the fireplace opening, reduce its need for air. It should be noted, however, that ready made outside air returns are often too small to supply enough air for a fireplace. The size of the damper opening is the size your air returns may need to be in order to effectively eliminate smoking.

Also note that outside air returns have clearance and location constraints: they need at least 1 inch clearance to any combustible material within 5 feet of the outlet (the fireplace) , and they should not be in the firebox itself, nor should they draw their air from any interior space such as the basement or garage, since drawing air from the building envelope instead of outside tends to depressurize the house and is therefore counterproductive in dealing with the smoking fireplace. There are also situations when a flow reversal might occur in an outside air return duct, leading to the possibility of sparks and embers traveling to the inlet.

Mechanical Problems of Stoves

After many years in the chimney sweeping business, I've concluded that most chronic stove problems, be they smokiness or excessive creosote buildup, are actually chronic operator problems: putting too much wood in the firebox at a time, not giving the fire enough air, etc. But there are troublesome mechanical problems that afflict stoves, too. I assume in this discussion that the chimney and smokepipe are clean and that the stove is not defective. The things to check are:

The chimney flue. If it's too large in diameter, too short or too cold, or a combination of the three (a situation common to new stoves installed in fireplace flues), the system may not provide a strong enough draft for the stove, with excessive creosote buildup and smoky operation the result. The solution is to furnish the stove with an appropriate chimney, either by lining and insulating the existing one or by replacing it.

The stovepipe. If it's too long or has too many elbows (more than two), it may allow the smoke to cool too much, which could reduce the effective draft. Stovepipe smaller in diameter than the stove's collar is also an absolute no-no. The solution is to install properly sized pipe in a less restrictive configuration, which is often easier said than done.

"Stack effect." In some situations, particularly a tight house with a stove (or fireplace) in the basement or on the ground floor, the house can become a more efficient chimney than the chimney itself, causing the smoke to flow down the chimney and into the house. This will most often happen when the stove is burning slowly and there are windows or doors open, or exhaust fans running, on the upper levels of the house but not the lower. What happens is that the warm air in the lower part of the house rises in compliance with the well-known physical law, its buoyancy abetted by the draft created by the open windows and bathroom fans. This causes low pressure to develop at the bottom of the house—lower than the pressure at the bottom of the chimney—and, presto, your house becomes a smoky chimney. To determine if stack effect is your problem, close any windows, doors and fans (don't forget the exhaust hood on your kitchen range) on the upper levels of the house, and open a window on the lower level. If the smoking stops, stack effect is the culprit. You may be able to solve the

problem by not opening upstairs windows and avoiding slow fires, but more likely will have to provide a source of outside air for the stove.

Sometimes two or more problems will coexist in the same star-crossed system. A case in point is a system I encountered recently. A new fireplace stove had been installed in a one-storey addition to a two-storey house. The stove smoked badly when initially fired up in cold weather, but worked perfectly in milder conditions. Also, the hearth was very cold to the touch, usually a dead giveaway that a cold chimney is the problem. The flaming torch over the damper helped, but the problem didn't really disappear until I tried shutting the doors between the addition and the rest of the house and opening a window near the fireplace stove, suggesting that stack effect was involved, too.

Sometimes an obvious chimney problem isn't a problem, as I once discovered. I was asked by a mason to help him determine why a new fireplace always smoked. It was a classic setup for stack effect: the smoky fireplace was on the ground floor, and the fireplace on the floor above didn't smoke. The house had an open floor plan, was three storeys tall, and tightly constructed. I suggested all of the cures for stack effect, and went on my way, confident as could be. A few days later the mason called again; my fixes hadn't worked at all. Back I went, puzzled. This time I got on the roof, as I should have done initially and hadn't, figuring that the chimney was too new to be dirty. What I found was a big blob of mortar at an offset in the flue. It reduced the cross-sectional area of the flue by at least half, and accounted entirely for the smoky behavior of the fireplace. Embarrassing, but instructive.

Back-Puffing. Your stove may back-puff occasionally for reasons that are not wind related. You'll recognize a back-puff because it is sudden, sharp and often followed by several more. What's happening is that the wood burning in the stove is producing combustible gases, but the oxygen supply is insufficient for them to ignite or exit the flue. As the stove draws more air, it provides enough for the stationary gases to suddenly ignite, which they do, with an explosion that pushes smoke out through the stove's air vents. The vacuum created by the explosion pulls in more oxygen, causing another explosion if there's enough gas available. The management of the stove is what causes back-puffing—specifically, too abruptly and completely restricting the oxygen supply to a large, hot fire. The combustion is hot enough to produce gas, but there isn't enough oxygen to burn it. The solution? Give the stove more air.

What to do in a Chimney Fire

A chimney fire is not so much a performance problem as it is the culmination of several—generally operator-induced—performance problems. Because a homeowner's initial reaction to a chimney fire may well determine how destructive it becomes, it is well worth quickly reviewing the correct procedure to follow.

Detection of a chimney fire is not usually a problem. It will likely announce itself with a prolonged roaring noise, smoke and odor in the house and thick, dark smoke and/or sparks and flames coming out of the top of the chimney. Some chimney fires are not so dramatic, probably because they haven't enough fuel or oxygen to really take off, but all chimney fires are potentially destructive and should be taken seriously. To people who regard them as a harmless way to clean a chimney, I can only say that physicians used to bleed people who were ill, too; all of the available objective evidence indicates that both practices are foolhardy.

For people with chimney fire paranoia, there are detectors available that indicate when the flue temperature is conducive to a fire.

Should you have a chimney fire, the first two things to do, simultaneously if possible, are:

1. *Shut off the fire's air supply* as completely as possible by closing all air intakes on a stove or covering the opening of a fireplace with something nonflammable.
2. *Call the fire department.* Many people don't take this step, probably out of embarrassment or to spare the firefighters, but I've been assured by every firefighter I've ever asked that they want to be called. It's much easier to contain a chimney fire than it is to extinguish a structure fire. If you are alone when the fire occurs and can't take Steps 1 and 2 simultaneously, I believe that you should shut off the air supply before making the call, unless the fire has already spread to the house or cutting off the fire's air will take you too long. The reason is that depriving the fire of oxygen in the beginning may be the critical factor in keeping the flue temperatures from skyrocketing and

destroying the chimney. Never open the stove door or cleanout after shutting off the air. To do so is to supply the fire with a sudden abundance of oxygen and could result in an explosion.

When you have shut off air to a chimney fire and called for help, the final step is simple but hard: prepare to evacuate the building, should that become necessary.

I've recently heard an alternative theory concerning dealing with a chimney fire: call the fire department but do not shut off the air. The logic behind this procedure is that many chimney fires handled the way I suggest above will smolder for hours and, in so doing, will transmit more heat to the structure than they would if allowed to burn out quickly. Some will even regain intensity after the firemen have left. While there may be some good evidence supporting this theory, I'd wait until it is generally accepted by fire marshals, fire investigators and chimney sweeps before practicing it. I'd also want to be absolutely certain that my chimney was sound before I let it burn unimpeded.

After a chimney fire, do not use your chimney until it has been thoroughly inspected by a competent chimney sweep.

Water Problems

Again, not a problem with performance, but a potentially damaging problem, nonetheless. Water, added to the residues of combustion found even in "clean" chimneys, can speed the deterioration of masonry and metal flues. Water penetrating a masonry chimney, either through the crown or the porous brick or block exterior, can result in significant damage, particularly when it freezes. Not all chimneys seem prone to either problem, but if you notice water in your system—it may show up in the cleanout or the fireplace—or see signs of water damage such as rusty metal components or flaky, crumbling masonry, it behooves you to take remedial action.

Water entering the flue is usually easily stopped by installing a cap (a removable one, please, to keep your sweep happy). Water that seems to be penetrating the chimney's exterior may be coming through the crown, the masonry walls, or the flashing. If the crown is cracked, have it repaired or replaced. If the flashing isn't securely sealed, repair it by caulking it with silicone, or replace it if necessary. If these steps don't cause the problem to abate, you'll need to consider waterproofing the chimney itself, a relatively inexpensive service offered by many chimney sweeps, and something that many people can do for themselves for the price of a can or two of masonry sealer.

Water-related chimney problems can be vexing to diagnose and solve. It may make sense and save you money in the long run to get professional help if your chimney has sprung a leak and the source is not obvious.

CHAPTER 8

The Woodpile

Deciding to heat with wood is one thing, but finding a reliable source of fuel can be quite another. The first question to ask yourself is whether you should buy your wood cut, split and delivered or do some or all of the work yourself.

The primary advantage of buying ready-to-burn wood is convenience: processing firewood is time-consuming, even for skilled and well-equipped professionals. The work also requires a fair amount of physical strength and stamina, has inherent dangers and is best performed with tools and vehicles that are expensive and that are not found in the typical homeowner's tool chest or driveway. Another advantage is that it is easier to find firewood dealers in many areas than it is to find available standing timber.

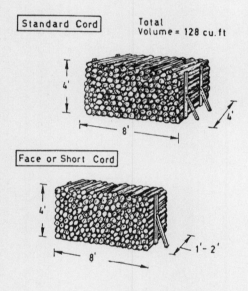

The disadvantage of buying ready-to-burn (or, at least, ready-to-pile) wood is, of course, expense. The dealer must cope with all of the aforementioned problems inherent in wood harvesting just as surely as you would, and his coping has a price. In my neck of the woods, for example, a cord of unseasoned hardwood (*i.e.*, a stacked pile measuring 8- by 4- by 4-feet and containing 128 cubic feet, including an acceptable and inevitable 30 to 40 cubic feet of air space), cut, split and delivered, will cost $75-$85. The same cord cut, but not split, will cost $60-$65, if you can find anyone selling it that way. That cord uncut and unsplit—delivered to your house in random log lengths of 8 feet to 18 feet—will probably cost $50. The biggest savings by far come when you do the whole job yourself, starting with standing trees, even if you don't own your own woodlot. The stumpage price (the price the woodcutter pays the landowner) for firewood is usually $5 or $6 per cord. Why the big difference in price for wood in its various forms? Let's look at the firewood business.

Firewood Economics

A friend of mine who sells some firewood ("but not for a living, fortunately") thinks that somebody ought to do an MBA thesis on firewood prices and the market forces that affect them. He says, and I agree, that in our area (and probably in most well-forested areas) the commodity is seriously underpriced, whether you consider its heat value (BTU's per dollar) or the money that an entrepreneur can make per hour selling it. The price of firewood has, in fact, changed little in the past twenty years. Adjusted for inflation, the price has certainly dropped, while the price of the tools and fuel needed to harvest it has gone up along with everything else. In 1974 my top-of-the-line chain saw cost $290, gas cost 50 cents per gallon, and firewood cost $65-$75 per green cord, cut, split and delivered. The top-of-the-line saw I own today cost $850, gas costs $1.60 per gallon here in Vermont (God help the poor Canadians!), and firewood costs $75-$85 per cord, cut, split and delivered. Why? And why can you save $70-$80 per cord by performing the whole harvesting operation, but only $25 per cord by bucking and splitting it yourself?

The answers are competition and equipment. In forested areas there tend to be a fairly large number of people who sell firewood, whether on a full-time basis or part-time as a by-product of logging, which also tends to be a common activity in such areas. This competition keeps the top price relatively low, particularly because loggers and other cutters who don't depend upon firewood as their main salable commodity will often undersell everyone who does and thus force the price down. Another factor keeping the price low is that if wood buyers had to pay a per-cord price that realistically reflected wood's value, many would switch to alternative fuels which, though more expensive, require no labor from the consumer.

The equipment factor explains why the savings per cord aren't substantial until you do all of the work yourself. Chain saws and splitters don't cost a fortune, but skidders and loader-equipped log trucks do, and that's the equipment it took to make that six-cord pile of log-length firewood appear at the end of your driveway. That's why it cost $50 per cord, uncut and unsplit. You still have to perform the most arduous and time-consuming labor to get the wood ready for your stove or fireplace, but you won't need $150,000 worth of machinery to do it.

Getting your Wood

Perhaps the best way to help you decide which wood procurement method most suits you is to describe four scenarios including comparative cost and time estimates. We'll assume that all four scenarios are at least theoretically available to you. (Clearly, city dwellers would have a hard time arranging for delivery of log truck loads of uncut firewood, and an even harder time convincing their neighbors not to complain about the chain saw noise. Clearly, too, many people, for various reasons, cannot consider cutting standing trees.) The prices for green wood that I use are fairly accurate in the rural northeastern U.S.; if they are higher or lower where you live, use the correct figures. Also, the times given for the tasks described are subject to enormous variation: a logger with a 90cc chain saw bucking large-diameter wood into 4-foot lengths can make a cord faster than you can drink a cup of hot coffee. A weekend cutter with a dull 35cc saw bucking small-diameter wood into 14-inch lengths will take a lot longer—almost forever. Similarly, splitting a cord of straight-grained ash might take less than the hour and a half that I allow; splitting acord of American elm, especially by hand, will likely take much longer, if you're desperate enough to do it. Keeping these variables in mind, let's survey our four scenarios.

1. *You buy your wood cut, split and delivered.* The cost will be $75-$85 per cord. The only labor left for you to perform is piling the wood so that it will season properly. This job requires no tools and will probably take an hour or less per cord unless you have to move the wood some distance from where it's delivered to where it is to be piled, in

which case you'll need a wheelbarrow and probably a good deal more than an hour, depending upon how far the wood must be moved. At some point during the wheelbarrowing, you will probably figure out a way of having the place where the wood is delivered next year be the place where it is to be piled.

2. *You buy your wood cut and delivered, but unsplit.* The cost, if it's available, will probably be $60-$65 per cord. In addition to the labor involved in piling, you will need to do the splitting. The tools required might be a $25 splitting maul or a $2000 hydraulic splitter, depending upon your bank balance and your proclivities. The time involved? A common old-timer claim is that a good worker should be able to split a cord per hour with a splitting maul. Well, maybe there aren't any good workers left, and maybe the truth has enjoyed a bit of embellishment over the years. It's also possible that old stoves and furnaces accepted large and unsplit chunks more gracefully than do modern stoves, so fewer sticks per cord actually got split in the old days. Possibly, too, with fewer foresters around, more straight-grained, forest-grown trees found their way to the woodpile, making for easier splitting. Splitting a cord of wood may well be a tougher job now than it was in our grandparents' time. In any case it's realistic to allow 1½-2 hours per cord if you're splitting by hand and 1½ hours per cord if you're using a hydraulic splitter. Your savings compared to buying your wood cut, split and delivered will be $10-$25 per cord, and the additional time required will be 1½-2 hours, so by splitting your own wood you can give yourself employment that pays you anywhere from $5 to $17 per hour, though I wouldn't count on making the $17 very often.

3. *You buy a log truck load of uncut, unsplit, delivered firewood.* The price will probably be $325-$375 for the load, and the load should contain 6-8 cords. Being in some ways conservative, we'll say that your price per cord will be $50. In addition to splitting and piling this wood, you will have to buck it—that is, cut it to length. For this you will need a chain saw or a bow saw. The bucking time I allow assumes that you are using a chain saw. A bow saw will take much longer. Without getting entangled in the variables (muddy logs dull your chain—you spend an hour filing for every cord you cut; your saw is small and slow; your saw is big and fast, etc.), we'll allow 1-2 hours for bucking

a cord of average logs into 18-inch stove lengths. You've worked 2½-4 hours more than you would have if you'd bought cut, split and delivered, and you've saved $25-$35 per cord. Not a great hourly wage, but few woodcutters get rich, and unlike most jobs, this one pays more in direct proportion to your production as it increases with experience.

4. *You buy stumpage* (the right to cut a certain number of standing trees on someone else's property). This should cost $5-$10 per cord. In addition to bucking, splitting and piling, you will have to fell, limb, skid (move the wood from the stump to the truck) and truck your wood. The necessary tools vary depending upon where the trees are, but at a minimum would include a saw and a pickup truck, and at the other extreme would include a good portion of the $150,000 worth of equipment to which we referred earlier. Before calculating your potential savings, let me point out the obvious: if you don't already have a truck, it probably won't pay you to buy one just to haul firewood. You'll wear it out and replace it before it amortizes itself.

If you'll bear with me through a little oversimplification, I'll show you what I mean. Say that you save $80 a cord by buying stumpage at $5, and you burn 4 cords per year. Without allowing for your time, equipment and increased wear and tear on the truck, you save $320 per year over the cost of wood cut, split and delivered at $85 per cord. In ten years, that's only $3,200 saved, and you'd have to be lucky indeed to find a $3,200 truck that would give you ten years of service. The truck looks like an even worse bargain when you consider that it will only save you $180 per year (for four cords) over the cost of having log truck loads delivered to your home.

If you already own a truck, can you save money by doing the whole firewood operation yourself, soup to nuts? Perhaps. As with any task related to wood harvesting, the variables are numerous. How far from home to the trees? Can you drive up to them, or will you have to haul them to the truck somehow? How far? Are you a reasonably skillful woods worker, or a neophyte? And the list could go on. A few summers ago, I bought firewood stumpage for $6 per cord in the state forest. The woodlot was 15 minutes from my house, and the trees marked for cutting were mostly within 100 feet of truck access, with many closer. My truck held a little less than half a cord, split and stacked, with room left for tools, and I was able to fill it in a little less than an hour and a half. Add the driving time and the hour per cord to pile it, and you find that I was processing wood from tree to pile at the rate of five hours per cord. Subtract the $6 stumpage fee, and you find that I was making $10-$13 per hour, before figuring in gas, oil and depreciation on the truck and saws. I would do it again because I enjoy it, I already have the equipment, training and experience for the job, and the wood does represent a several-hundred-dollar cash outlay that I don't have to make, but it certainly doesn't represent a windfall savings over having the wood delivered in some form.

There is a good deal of firewood available that is half a mile or more from the nearest access. Even if you could get this wood for free, it would never pay you to carry it to your truck (unless someone bought the movie rights from you) or to buy or hire a skidder to do it. My point is that once you get away from having wood delivered to you, all bets are off; sometimes harvesting your own wood will be worthwhile, particularly if you factor in some intangibles such as fresh air, fun and exercise, but often it will be a losing proposition unless you own a woodlot and have a truck, tractor and/or draft animals.

The math and advice I've just presented seem to point to the wisdom of buying wood fully processed, but having done my duty by making that clear, I will now encourage you, if your temperament, circumstances and physical abilities permit, to consider doing some or all of the processing yourself. Intangibles shouldn't be discounted, and a true understanding of the work that went into your winter's fuel might make its warmth that much more comforting. But do keep a healthy dose of clear-eyed evaluation handy, unless you have more time to spend than you can use.

If you resist rising to this bait, read on: in the next section we'll consider how to buy firewood fully processed.

Buying Firewood

As a chimney sweep, I get to know my customers' chimneys, stoves and fireplaces and, sometimes, their woodpiles as well. One day, while cleaning a regular client's chimney, I noted a small accumulation of dangerous glazed creosote—new to this chimney and very unwelcome, since it is difficult to remove and burns intensely in a chimney fire.

"It must be that wood I bought," said my client when I gave her my report. "The fellow said it was dry, but it sizzles a lot. You look at the pile and see what you think."

I did look, and found wood which, if not cut yesterday, was surely not seasoned; even in the dim light of the woodshed it was easy to see the fresh color of the sapwood.

Later that same day I had just finished sweeping a first-time customer's chimney when his wood was delivered: three supposedly seasoned cords for $375. Before the truck had even pulled into the driveway, I caught the fresh, sappy aroma of green wood. A quick look at the pile confirmed what my nose had already told me: this expensive "seasoned" wood was freshly split and likely freshly cut as well.

Buying firewood is tricky. The business is unregulated; there are no state or federal inspectors, no trustworthy stamps of approval. The quality and even the quantity of the product are difficult to assess. It's easy enough to measure a neatly stacked cord, but can you recognize 128 cubic feet of wood dumped in a heap at the end of your driveway? Can you tell dry wood from green? Both of my aforementioned customers are relatively knowledgeable woodburners, and both, nevertheless, had to make do with green wood after paying for seasoned. Some firewood dealers are highly competent and scrupulously honest, but some are neither, so let the buyer beware! As a consumer, you must be well informed or very lucky if you are going to consistently enjoy the warmth and satisfaction that good wood can provide.

You can avoid the uncertainty of the market by cutting your own wood, but this course substitutes its own problems, previously discussed, for the ones it eliminates and is probably a viable alternative mainly for those who enjoy it. If you don't find yourself in that relatively select group, you're wise to make a smart shopper of yourself.

Is It Seasoned, Dry, Wet or Green?

Make no mistake, it is better to burn dry, seasoned wood than green, wet wood. We've probably all heard people laud the virtues of mixing green and seasoned woods for a lasting fire, and perhaps it worked well in old, nonairtight stoves. But modern airtight stoves, whether high-efficiency, catalytic or '70's vintage black boxes, need high temperature fires. Burning green wood in an airtight will almost certainly result in low heat output, heavy glazed creosote accumulation in the venting system and, possibly, dangerous chimney fires. Wood, as it seasons, doesn't lose creosote; it only loses water. When green wood is burned, much of the heat it produces is used to evaporate the excess water instead of warming your house and burning off the tars and creosote produced by the fire.

The words "seasoned" and "dry" are generally used synonymously when describing firewood that has been cut, split and piled for one year. Wood that was cut a year ago and split yesterday shouldn't be called seasoned. You need to question the dealer closely about his seasoned wood before you buy it. When was the wood cut? When was it bucked? When was it split? Was it piled off the ground? Under cover? Off the ground and under cover are not part of the generally accepted definition of seasoned wood, but wood split six months ago and stored that way may well be drier than wood split a year ago and left on the ground exposed to the elements.

What if the dealer lies to you? Then it pays to be able to recognize wood that is seasoned and wood that is not. None of the following indications are foolproof—radial checking, for example, can occur within a few weeks of cutting—but these, coupled with experience and the slightly jaundiced eye that experience often brings, should enable you to get what you are paying for most of the time.

SIGNS THAT WOOD IS SEASONED

Weight. Seasoned wood is much lighter than green wood of the same species.

Smell. Green wood often has a pleasant, sappy aroma (or, in the case of some species such as red oak and elm, an unpleasant, sappy aroma). Seasoned wood will smell like wood, but not as strongly.

Loose bark. As wood dries, the bark adheres less tenaciously. This does not mean that any wood with its bark firmly in place is green, but out of a cord of seasoned wood, you should notice a modest amount of barkless wood and woodless bark.

Color. The sapwood visible on the ends and split sides of wood billets fades as wood seasons. Different species have different shades and colors, obviously, but if your new load of wood is bright and fresh in color rather than dull and subdued, you'll want to take a closer look.

Radial cracking. As wood dries, it often develops cracks, or *checks*, that radiate from the heartwood out toward the bark, and are visible on the cut ends of the stick. As mentioned before, this phenomenon may be evident long before the wood is dry, but if it is not evident at all, this is another reason for a closer look.

Green cambium. If you're pretty certain that the wood in question is green, peel a little bark back with a sharp knife and check the *cambium*, which is the very thin layer between the bark and the sapwood. If it is green, so is the wood.

BUY GREEN

While it certainly is necessary to burn dry wood, it is not necessary to buy dry wood if you are organized enough to buy your wood a year or so before you need to burn it. There are three compelling advantages to this strategy:

Certainty. If you store it properly yourself, you know that your wood is seasoned.

Price. Green wood typically costs anywhere from $15 to $50 less per cord than seasoned.

Supply. Seasoned wood is not always available in many areas, particularly as winter approaches, so drying your own could save you a miserable heating season. Since you have to store seasoned wood in exactly the same way as you would store green, there is, in fact, no adavantage to buying seasoned.

QUANTITY

An even more common complaint among firewood buyers than supposedly seasoned wood that turns out to be green concerns the cord that is not a cord. You might, of course, buy your wood by the "face cord" (8 feet by 4 feet by whatever length you agree upon), the half cord, or in lots of more than a cord, but, in any case, the price of wood is mainly determined by volume, and unless you are paying the dealer extra to stack the wood or he or she brings it stacked in the truck for your inspection (laudable but unusual), you'll either have to accept the dealer's word regarding quantity or learn to recognize a cord (or fractions and multiples thereof) when it is dumped in a heap. A good, though strenuous and somewhat imprecise way of doing this is to stack and measure a cord (or whatever quantity you have delivered), throw it into a pile and familiarize yourself with the look and approximate dimensions of the pile.

Another strategy for estimating firewood quantity involves evaluating the truck in which the wood is delivered. How much will it hold? Some years ago I ordered two cords of sugar maple. The dealer explained that the wood was not cut to length or split—hence the low price that had attracted me in the first place—but was cut to the size he could handle: large-diameter sticks might be 18 inches long, and small-diameter sticks might be 8 feet long. He said he could bring a cord at a time.

"What have you got for a truck?" I asked, knowing that a cord of green maple might weigh 3 tons or more. (A cord of dry might exceed 2 tons.)

"Half-ton Chevy with built-up sides," he answered.

"We'll see," I said to myself. Besides being skeptical that his truck could actually carry the weight of a green cord of sugar maple, I knew that random length and width wood thrown—not stacked—on the truck would contain an extraordinary volume of air, making it unlikely that the advertised two cords would be for real. To make a long story short, the two loads measured 1¼ cords bucked, split and piled. Fortunately, the dealer was honest enough to make up the difference.

A truck will hold significantly more if the wood is stacked. For example, a dumptruck I used to use to deliver wood held one cord dumped on with a front-end loader, but 1¼ cords stacked. A truck cargo area that measures 128 cubic feet won't contain a cord unless the wood is tightly stacked or is heaped well above the sideboards.

QUALITY

Getting the amount of wood that you pay for is certainly important, but don't let concern about volume distract your attention from the reason you're buying wood: its heat value, not the air it will displace in your woodshed. For example, a cord of shagbark hickory takes up the same space as does a cord of aspen ("popple" if you don't live in town). The hickory, however, weighs about twice as much as the popple and has about twice as many available BTU's. A half cord of hickory, therefore, is worth as much as a full cord of aspen. Before you can decide which dealer is offering the best price, you obviously need to know what kinds of wood he's selling. "All hardwood" is not a good enough answer, and neither is "mixed hardwood." Apple is hardwood, and so is elm, but recall from the table in Chapter 2 that a cord of apple is the equivalent of 244 gallons of heating oil, while American elm only provides the BTU's of 176 gallons. "Maple" is not a good enough answer, either, at least in areas that have both sugar maple and soft maple (*i.e.*, any variety of maple other than sugar maple and black maple), since sugar maple is the equivalent of up to 44 more gallons of oil per cord than are some soft maples.

In fairness to firewood dealers, I should point out that unless you're paying a premium for quality, a fair cord is a mix of the firewood species commonly found in your area. You shouldn't expect every stick to be top of the line, but neither should every stick be poor to mediocre unless the price is reduced.

Here again, you will do well to become an informed consumer. It's not as easy to identify wood when it's cut and split as it is when it's part of a standing tree, replete with leaves, buds and twigs, but it can be done. Some years ago a firewood customer ordered two cords of ash. I delivered it: two full cords, and every single stick was ash. After he'd had a chance to inspect it, he called.

"I ordered ash. What you brought is sugar maple, isn't it?" Sugar maple would, in fact, have been a better buy, but I reassured him that his wood was ash, and he apparently burned it happily. It seems possible, though, that I could have brought him popple, which certainly isn't as good a buy as ash. He was properly suspicious, but not properly informed.

Books, knowledgeable friends and experience will help you master firewood identification. You'll still be fooled occasionally; even experts are. (An aquaintance of mine once successfully passed red elm off as ash to a sawmill.) But if you know your trees, you'll be much better able to get your money's worth when buying firewood.

Heating your house satisfactorily with wood requires more work and knowledge on your part than does any other common type of residential heating. Few are the consumers discerning enough to notice a qualitative difference between one tank of oil or gas and another, but equally few are the consumers who wouldn't notice the difference in performance between good wood and bad. The challenge is to learn enough so that you can identify good wood before you try to burn it and before you pay for it. If you care about these things and enjoy working toward self-reliance, becoming knowledgeable should prove to be a pleasure in and of itself, a pleasure at least matched by the reliable warmth and comfort of your good wood.

Storing and Seasoning Firewood

Whether you've paid a premium for cut, split and delivered firewood, or you've labored long and hard to cut, split and deliver it to yourself, you still have a critically important task ahead of you: storing your wood. A pile of firewood is full of potential heat which may be realized when you feed it to the flames in your stove, furnace or fireplace. That potential may also go unrealized—at least in part—if you neglect the relatively simple chore of properly piling and protecting your fuel. I say simple because you don't have to hide your wood from two of the elements—wind and sun—as they will help it dry. Water is the element to guard against, your woodpile's main enemy. Piling your wood under dripping eaves or in an uncovered heap in the yard is comparable to buying or growing a bunch of beautiful fresh asparagus and then over-cooking it: you get no refund for ruining it.

Green wood that is piled for seasoning will dry slower than necessary or not at all, depending upon how much moisture it's exposed to, if it isn't protected. Seasoned wood left unprotected will become, for all intents and purposes, unseasoned as it regains the moisture it lost. Don't believe me? Take a stick of bone-dry firewood and submerge it in a tub of water for a few days. Now try to burn it. For that matter, toss a stick of dry wood into some tall grass and then try to burn it after a good two-day rainstorm: it'll be awhile before it produces as much heat as it does sizzle.

Here are three simple rules for good wood storage:

1. *Allow air circulation.* You can accomplish this by leaving the sides of the woodpile uncovered and by not stacking your wood so tightly that the pile looks like a well-made jigsaw puzzle. If you have the room, piling your wood in long, narrow tiers—one stick wide, ideally—that run north-south to take full advantage of east and west

winds will maximize air circulation.   Norm Hudson of the Vermont Department of Forests and Parks says un-equivocally that siting your woodpile properly for air circulation is the

most important factor in wood storage. "Pick the best place to hang your laundry—dry and windy," he says, "and that's where to pile your wood. I like my woodpiles one tier wide, as long as they have to be, and as high as my wife can comfortably reach."

2. *Protect your wood from rain and snow* by covering the top of the pile. You can do this with a tarp or a woodshed roof, but in either case, it's best if the cover isn't resting on the top of the woodpile: leaving a few inches of air space there abets circulation. The cover or roof should also extend be-yond the wood in all

directions to keep at least some wind-driven precipitation off the pile. How far the cover extends depends upon how windy it is where you live and how high you stack your wood (how tall your wife is?).

3. *Pile your wood off the ground.* Water attacks from below as well as from above, so if you don't provide something other than bare ground for your wood to rest on, the bottom layer of your pile will stay wet. This may not seem like a big deal, but if your pile is four feet tall, you might be losing 10% or more of your wood to ground moisture. You paid just as much or worked just as hard for the wood on the bottom as for the wood on the top, so why waste it? I find wooden pallets—preferably two layers of them—ideal as the floor under my woodpile: They keep the wood well off of the ground, and because they're slatted and hollow, they allow air to circulate from the bot-tom. It's almost too good to be true.

A note about location: On the one hand, you'd like to pile your wood as near to the stove as possible. On the other hand, the termites and carpenter ants would probably like that as well, or better, than you do. Pest control experts I've consulted say that the farther from your house you can pile the wood, the better—as much as 150 feet isn't too far, according to at least one, but it's probably too far to carry it—once the kids have left home, at least. My woodshed is about 50 feet from my house. The shed has carpenter ants, but the house doesn't. I've known people who kept their wood in their basements or in attached woodsheds and had no trouble with insects, and I've known other people who have gotten so tired of hiring exterminators that they've built new woodsheds well removed from their houses. So the location of your woodpile is a compromise between your fear of bugs and your fear of carrying armloads of wood great distances in inclement weather. So much about woodburning seems to be a matter of compromise.

STACKING

Why bother stacking wood? As chores go it's fairly time-consuming, moderately arduous and thoroughly unexciting, and I suppose, with a tip of my hat to the lazy side of the brain, that it wouldn't be necessary if you had unlimited appropriate—covered and off the ground—space in which to toss your firewood.  Most people do not have such capacious woodsheds, however, so a good reason for stacking wood is that stacked wood of a certain quantity takes up less space—or, rather, has a smaller footprint—than does the same quantity of wood tossed in a heap. Other reasons are that stacked wood is more receptive to air and sun than is a big, dense pile, and nicely stacked wood looks neat and shipshape, if that matters to you.

How should you stack your wood? Again, this seems simple, but it's surprising how many otherwise capable and intelligent people are unable to stack their wood so that it doesn't keep falling down. Falling down is, of course, the one wrong thing that a stack of wood can do.

Wood stacks tumble for several reasons: they're not properly supported at the ends; they're built too high but not straight enough; and strong frost heaves topple them. You can support the ends of your wood stacks either by piling against stakes driven into the ground or spiked to the pallets or boards that you're piling on, or by cross-hatching or log-cabin piling the ends of each tier.

My woodpiles have a tendency to get crooked and unstable when I pile more than 4 feet high, which I habitually do because of space limitations. This instability is partly because I'm always in a hurry and, consequently, somewhat inattentive, and partly—I'm trying to let myself off the hook a bit here—because firewood is not uniform, and the higher you pile it, the more you must compensate for its asymmetry. Nonetheless, I generally get away with building 6- or 7-foot-tall stacks by forcing myself to slow down, check my work, and pile those highly individual sticks of wood so that they complement each other and produce a plumb, stable, symmetrical whole. Although stacking may be unexciting, it does occupy hands and mind well enough to make the time pass quickly.

I've had apparently stable woodpiles tumble in the spring when the frost comes out of the ground. This probably wouldn't happen if I were willing to wait until after the frost is gone to do that chore. Repiling two or three cords of wood that have tumbled together sideways into the tall, wet grass outside the shed is enough of a pain in the neck to make a reasonable person learn patience.

SEASONING WOOD

The time-honored way to season wood is to buck, split and stack it off the ground and under cover—that phrase again—for one full year. This method will yield firewood with a moisture content of 15-20%. (Green wood may have a moisture content of 50%.) It's well worth noting, however, that drier is not always better: wood can be too dry. As we mentioned earlier, wood does not lose any creosote as it dries; it only loses moisture. What would happen if you dried your wood in, say, a greenhouse or even a covered shed for two or three years? Probably the moisture content would dip well below 15-20%, and the wood would ignite and burn very easily. In an open fireplace or, perhaps, a masonry heater, this might be no problem at all. In a woodstove or furnace, however, this super-dry wood would throw so much heat so quickly that you would shut down the air supply in order to avoid damaging the heating system and cooking yourself. The product, in the firebox, of high temperatures and little air is smoke—lots of it. Smoke is creosote, and that's just what you'd get. This is why the National Chimney Sweep Guild recommends using wood with a moisture content of 15-20%.

So, if your wood is properly air-dried for one year, you should be all set. But what if you don't have a year? What if you don't get your wood until May or June? You're still all set, provided that you store it properly, because—and here's where the cat jumps out of the bag and mauls a cherished piece of folklore—according to a field study conducted by the U.S. Forest Service, wood properly piled in a dry and windy place will reach a moisture content of 20% in as little as two weeks! If rain falls on it, of course, the moisture content increases. If it's piled in a damp area where little wind reaches it, the wood may never reach 20% moisture content. So we're back to proper storage: if you store your wood properly in the right place, you don't need a year to season it. Of course, there are variables, such as weather, over which you have no control. If it's rainy while you are hoping to shortcut the seasoning process, all bets are off, so planning to get your wood ready a year before you need to burn it is still the best policy. But, if conditions are right, you can have burnable wood in far less time.

OTHER SHORTCUTS

Over the years people have devised several ways of seasoning wood quickly, at least in theory. In light of the information cited above, it seems clear that you needn't take heroic measures—how much sooner than two weeks hence do you need seasoned wood?—but in case, for some reason, you can't store your wood in a place dry and breezy enough for quick seasoning, here are three popular shortcuts. I must confess that, distrusting get-rich-quick schemes as I do, I haven't tried any of these methods, but all three have enthusiastic proponents.

1. *The solar wood drier* is a variation on standard proper seasoning. In this case the sun, rather than the wind, is the key element. You stack the wood east-west so that the long side of the pile is facing south, and build a frame over it. The frame should be higher than the stack of wood and extend beyond it. Cover the frame with heavyweight black plastic, securely anchored, and let the heat of the sun cook the moisture out of your firewood. It is important that the plastic not be touching the wood and that the sides of the stack be open; otherwise

the water from the wood will condense on the plastic and drip back onto the wood instead of evaporating or dripping onto the ground. People who have tried the solar drier say that they get well-seasoned wood in two or three summer months.

2. *Fell trees in the spring.* If you make your own firewood from standing trees, you can try cutting them in the spring as the new leaves are popping out and then leaving them, unlimbed and unbucked but severed from the stump, until the new leaves are completely wilted. The theory is that the leaves, unable to draw moisture from the root system, draw it from the trunk of the tree, leaving you with dry wood. Drying wood this way supposedly doesn't take more than a few weeks. The theory is plausible, and I've been told by folks who've tried it that it works. My comments are that it doesn't really save much seasoning time over properly storing your wood (which you'll need to do anyway, after you go back and cut it up) and that, while it might work well on some woodlots, on others you'd have an awful mess of tangled trees to sort out when you did go back to buck, limb and haul.

3. *Kiln-drying firewood,* which is probably not of practical interest for a do-it-yourselfer, is a technique used by a few commercial firewood dealers. This method produces wood with a moisture content of 15-20% in two or three days.

A final note on the subject of seasoned wood: there's no denying that using properly dry wood is important, but, as any knowledgeable chimney sweep will tell you, having a woodburning appliance appropriate to your needs and operating it properly are just as critical, if not more critical, to the safety and success of your heating operation.

Good luck, and happy burning.

APPENDIX

Safety Checklist

In case you decided to skip the entire preceding book, here it is in a miniature nutshell; important matters to consider before—long before—your first autumn fire.

S T O V E

1. Clearances from all combustible materials.

2. Structural integrity: cracks, warping, functioning moving parts and catalytic converter.

3. Door gaskets and latches.

S T O V E P I P E

1. Clearances from all combustible materials.

2. Length and configuration of pipe conducive to safe and clean burning.

3. Three sheet-metal screws at each joint. Support brackets on long horizontal sections (4 feet or more).

4. Crimped ends pointing down.

W A L L   P A S S - T H R O U G H S

1. Clearances to all combustible materials.

2. Positive connections to chimney and stovepipe.

FIREPLACES

1. Structural integrity: mortar, firebrick, damper.

2. Screen/glass doors functional.

3. Clearances from all combustible materials: wooden trim and mantels at least 6 inches from fireplace opening; 12 inches if mantel or trim projects 1½ inches out from above opening. Hearth extension should extend 16 inches in front and 8 inches to the sides of fireplace opening less than 6 square feet. 20 inches to the front and 12 inches to the sides for larger openings.

CHIMNEY

1. Structural integrity: liner, walls, crown, mortared joints, thimbles, cleanout doors.

2. Cleanliness.

3. Separate flue for each appliance.

WOOD

1. Seasoned.

2. Stored properly.

ASH DISPOSAL

1. Assume ashes contain hot embers.

2. Store in tightly covered metal container.

3. Don't vacuum

4. Don't dump anywhere wind can reach and scatter hot ashes.

BIBLIOGRAPHY AND SOURCES

National Fire Protection Association. *NPFA 211 Standard for Chimneys, Fireplaces, Vents, and Solid Fuel-Burning Appliances,* 2000 Edition. Quincy, MA.

National Chimney Sweep Guild. *Successful Chimney Sweeping.* Olney, MD. 1987.

Dirk Thomas. *The Harrowsmith Country Life Guide to Wood Heat.* Camden House Publishing, Inc. Charlotte, VT. 1992.

Donald Culross Peattie. *A Natural History of Trees of Eastern and Central North America.* New York, NY: Bonanza Books. 1954.

John Gulland. *Reliable Chimney Venting Training Manual.* Hearth Education Foundation. Austin, TX. 1995

Natural Resources Canada. *A Guide to Residential Wood Heating.* Ottawa, Ontario. 1993.

Cooperative Extension Services of the Northeast States. *Wood as a Home Fuel.*

Jay W. Shelton. *Wood Heat Safety.* Garden Way Publishing. Charlotte, VT. 1979.

Daryle Thomas. *The Hearth and Cricket Shop.* East Wallingford, VT.

D. Cook. *The Ax Book: The Lore and Science of the Woodcutter.* Alan C. Hood & Co., Chambersburg, PA. 1999

Vrest Orton. *The Forgotten Art of Building a Good Fireplace.* Alan C. Hood & Co., Chambersburg, PA. 2000

INDEX